METHODS IN MOLECULAR BIOLOGY™

Series Editor
John M. Walker
School of Life Sciences
University of Hertfordshire
Hatfield, Hertfordshire, AL10 9AB, UK

For other titles published in this series, go to
www.springer.com/series/7651

Microarray Analysis of the Physical Genome

Methods and Protocols

Edited by

Jonathan R. Pollack

*Department of Pathology, Stanford University School of Medicine,
Stanford, CA, USA*

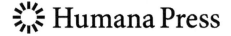

Editor
Jonathan R. Pollack
Department of Pathology
Stanford University
    School of Medicine
269 Campus Drive
Stanford CA 94305
USA
pollack1@stanford.edu

Series Editor
John M. Walker
University of Hertfordshire
Hatfield, Herts
UK

ISSN 1064-3745   e-ISSN 1940-6029
ISBN 978-1-60327-191-2   e-ISBN 978-1-60327-192-9
DOI 10.1007/978-1-60327-192-9
Springer Dordrecht Heidelberg London New York

Library of Congress Control Number: 2009926982

Printed on acid-free paper

Springer is part of Springer Science+Business Media (www.springer.com)

Preface

This volume in the Methods in Molecular Biology series focuses on microarray approaches and protocols to characterize the physical genome, with particular emphasis on higher eukaryote (animal) and cancer genomes. In the first years of microarray technology, efforts were directed mainly to profile expressed genes. In recent years, the microarray platform has been adapted to diverse applications directed to investigate the physical genome. This volume covers DNA microarray applications for the detection and characterization of genomic DNA-associated copy number alteration, loss of heterozygosity (LOH), cytosine methylation, protein-binding sites, regulatory elements, and replication timing. This collection will be of value to the molecular biologist or computational biologist interested in understanding the principles of these analyses, or in planning future experiments using microarrays to characterize the physical genome.

Contents

Contributors

MYLES BROWN • *Division of Molecular and Cellular Oncology, Harvard Medical School, Dana-Farber Cancer Institute, Boston, MA, USA*

JAMES BYRNES • *Cold Spring Harbor Laboratory, Cold Spring Harbor, NY, USA*

JENNIFER M. CAMPBELL • *British Columbia Cancer Research Centre, Vancouver, BC, Canada*

RAJ CHARI • *British Columbia Cancer Research Centre, Vancouver, BC, Canada*

BRADLEY P. COE • *British Columbia Cancer Research Centre, Vancouver, BC, Canada*

JOHN K. COWELL • *School of Medicine, Medical College of Georgia Cancer Center, Augusta, GA, USA*

GREGORY E. CRAWFORD • *Institute for Genome Sciences & Policy, Duke University, Durham, NC, USA*

DANIEL E. DEATHERAGE • *Human Cancer Genetics Program, The Ohio State University Comprehensive Cancer Center, The Ohio State University, Columbus, OH, USA*

ANINDYA DUTTA • *Department of Biochemistry and Molecular Genetics, University of Virginia, Charlottesville, VA, USA*

JÉRÔME EECKHOUTE • *Division of Molecular and Cellular Oncology, Harvard Medical School, Dana-Farber Cancer Institute, Boston, MA, USA*

TIM H.-M. HUANG • *Human Cancer Genetics Program, The Ohio State University Comprehensive Cancer Center, The Ohio State University, Columbus, OH, USA*

HANLEE JI • *Department of Medicine and Division of Oncology, Stanford University School of Medicine, Stanford, CA, USA*

NEERJA KARNANI • *Department of Biochemistry and Molecular Genetics, University of Virginia, Charlottesville, VA, USA*

YOUNG H. KIM • *Department of Pathology, Stanford University, Stanford, CA, USA*

WAN L. LAM • *British Columbia Cancer Research Centre, Vancouver, BC, Canada*

SHILI LIN • *Department of Statistics and the Mathematical Biosciences Institute, The Ohio State University, Columbus, OH, USA*

X. SHIRLEY LIU • *Department of Biostatistics and Computational Biology, Dana-Farber Cancer Institute, Harvard School of Public Health, Boston, MA, USA*

KEN C. LO • *Roche NimbleGen, Inc., Madison, WI, USA*

WILLIAM W. LOCKWOOD • *British Columbia Cancer Research Centre, Vancouver, BC, Canada*

ROBERT LUCITO • *Cold Spring Harbor Laboratory, Cold Spring Harbor, NY, USA*

MATHIEU LUPIEN • *Division of Molecular and Cellular Oncology, Harvard Medical School, Dana-Farber Cancer Institute, Boston, MA, USA*

CLIFFORD A. MEYER • *Department of Biostatistics and Computational Biology, Dana-Farber Cancer Institute, Harvard School of Public Health, Boston, MA, USA*

PAMELA L. PARIS • *Department of Urology, University of California at San Francisco, San Francisco, CA, USA*

JONATHAN R. POLLACK • *Department of Pathology, Stanford University, Stanford, CA, USA*

DUSTIN POTTER • *Human Cancer Genetics Program, The Ohio State University Comprehensive Cancer Center and the Mathematical Biosciences Institute, The Ohio State University, Columbus, OH, USA*

KEYAN SALARI • *Departments of Pathology and Genetics, Stanford University, Stanford, CA, USA*

YOICHIRO SHIBATA • *Institute for Genome Sciences & Policy, Duke University, Durham, NC, USA*

CHRISTOPHER M. TAYLOR • *Department of Biochemistry and Molecular Genetics, and Department of Computer Science, University of Virginia, Charlottesville, VA, USA*

EMILY A. VUCIC • *British Columbia Cancer Research Centre, Vancouver, BC, Canada*

PEI WANG • *Cancer Prevention Program, Division of Public Health Science, Fred Hutchinson Cancer Research Center, Seattle, WA, USA*

KATRINA WELCH • *Department of Medicine and Division of Oncology, Stanford University School of Medicine, Stanford, CA, USA*

IAN M. WILSON • *British Columbia Cancer Research Centre, Vancouver, BC, Canada*

PEARLLY S. YAN • *Human Cancer Genetics Program, The Ohio State University Comprehensive Cancer Center, The Ohio State University, Columbus, OH, USA*

Chapter 1

Introduction

Jonathan R. Pollack

Abstract

DNA microarray technology has revolutionized biological research by enabling genome-scale explorations. This chapter provides an overview of DNA microarray technology and its application to characterizing the physical genome, with a focus on cancer genomes. Specific areas discussed include investigations of DNA copy number alteration (and loss of heterozygosity), DNA methylation, DNA–protein (i.e., chromatin and transcription factor) interactions, DNA replication, and the integration of diverse genome-scale data types. Also provided is a perspective on recent advances and future directions in characterizing the physical genome.

Key words: DNA microarray, array CGH, SNP array, methylation array, ChIP-chip.

1. DNA Microarrays

DNA microarray technology emerged during the 1990s *(1, 2)*, made possible by large-scale DNA sequencing efforts that provided sequence information and DNA clones, together with advances in microfabrication techniques. DNA microarrays comprise thousands of DNA probes spotted or synthesized onto a solid surface in an ordered high-density array of rows and columns, and permit the highly parallel measurement of labeled target nucleic acids within a complex mixture. In the early years, spotted cDNA microarrays and Affymetrix GeneChip oligonucleotide arrays were the main platforms, but many different homemade and commercial platforms are now in use.

The first widespread application of DNA microarrays was the analysis of mRNA levels. Profiling gene expression in diseases like cancer provided new biological and clinical insights, including the discovery of new cancer subtypes *(3, 4)*, and signatures for

Jonathan R. Pollack (ed.), *Microarray Analysis of the Physical Genome: Methods and Protocols, vol. 556*
© Humana Press, a part of Springer Science+Business Media, LLC 2009
DOI 10.1007/978-1-60327-192-9_1 Springerprotocols.com

improved clinical outcome prediction *(5)*. More recently, DNA microarrays have been adapted and applied to characterizing various attributes of the DNA genome. This compilation of chapters provides information and protocols for using DNA microarrays to characterize the physical genome in relation to DNA copy number changes (and loss of heterozygosity), DNA methylation, DNA–protein interactions, and DNA replication, with an emphasis on higher eukaryote (animal) genomes, and in particular on the altered genomes of cancer.

2. DNA Copy Number Alterations

A cardinal feature of cancers is an aberrant genome characterized by gains and losses of chromosomes (i.e., aneuploidy) as well as more focal DNA amplifications and deletions *(6)*. Such DNA copy number alterations (CNAs) result in the altered dosage and expression of cancer genes. Indeed, identifying and mapping recurring CNAs has proven a powerful approach to discovering new cancer genes. Germline CNAs like microdeletions also play a role in the pathogenesis of developmental disorders *(7)*, and germline copy number variants (CNVs) may predispose to common diseases.

DNA microarrays provide a powerful tool to profile CNAs by comparative genomic hybridization (CGH) *(8)*. In array CGH, test (i.e., tumor) and reference (i.e., normal) genomic DNAs are differentially labeled and co-hybridized to an array containing DNA probes of known genome map position. Different types of DNA probes can be used, including large genomic DNA clones like bacterial artificial chromosomes (BACs), PCR-amplified genes/cDNAs, or 20–70-mer oligonucleotides. In general, BAC arrays provide greater measurement accuracy for each probe, while oligonucleotide arrays can be easier to manufacture and provide high mapping resolution by averaging measurements across neighboring probes.

In this volume, methods are provided for CGH using BAC arrays tiling the human genome (**Chapter 2**), as well as 70-mer oligonucleotide microarrays (**Chapter 3**). In addition, methods are provided for a variation of array CGH called representational oligonucleotide microarray analysis (ROMA) (**Chapter 4**). In ROMA, a lower-complexity representation (or sampling) of the genome is assayed, where the lower complexity can translate to higher hybridization signal intensities and measurement accuracy. A discussion of methods and analysis is also provided for CGH using commercial (Affymetrix) single-nucleotide polymorphism (SNP) arrays (**Chapter 5**). SNP arrays can be used to score not only DNA dosage but also loss of heterozygosity (LOH) (which can occur with DNA deletion and also with copy number neutral events like mitotic recombination) and ploidy. A novel approach for copy number and genotype determination is also detailed based on molecular inversion probes (MIPs) (**Chapter 6**), where a probe ligation step coupled with a barcode hybridization readout can provide increased assay sensitivity and specificity. CGH characterization of small specimens, like biopsies or microdissected material, can necessitate amplification of DNA prior to array analysis. Methods are provided to amplify genomic DNA while maintaining dosage representation (**Chapter 7**). Finally, statistical approaches are described to objectively call DNA gains and losses in array CGH data (**Chapter 8**).

3. DNA Methylation

DNA methylation is an epigenetic alteration associated with silenced gene expression. In mammals, DNA can be methylated at cytosines occurring as CG dinucleotides, often within "CpG islands" found at gene promoters. DNA methylation plays a key role in X chromosome inactivation and with imprinted genes where the paternally or maternally inherited allele is selectively silenced. In cancer, focal DNA methylation can silence tumor suppressor genes *(9)*, and identifying hypermethylated promoters can therefore implicate novel cancer genes.

In this volume, two methods are detailed to profile genomic DNA methylation using DNA microarrays. In the first approach, differential methylation analysis (DMA), methylation-sensitive endonucleases are used, where methylated DNA fragments protected from digestion are subsequently PCR-amplified, labeled, and hybridized to promoter (CpG island) arrays (**Chapter 9**). In the second approach, methylation analysis by DNA immunoprecipitation (MDIP), methylated DNAs are selectively immunoprecipitated, then PCR-amplified and analyzed by hybridization to genome-tiling arrays (**Chapter 10**).

4. DNA–Protein Interactions

While cells in different organs have the same DNA content, their distinct transcriptional programs and phenotypes are largely set by the constellation of chromatin proteins and transcription factors associated with the DNA. It is of great interest therefore to define the genome sites of DNA–protein interaction and to characterize differences occurring in disease states like cancer.

In this volume, two distinct approaches are detailed for microarray-based profiling of DNA–protein interactions. In the first, chromatin immunoprecipitation coupled with gene chips (microarrays) or "ChIP-chip" (**Chapter 11**), proteins are cross-linked to DNA, then bound DNAs are immunoprecipitated (using an antibody to the protein of interest), labeled, and hybridized to a promoter or tiling array. Analytical approaches are also described to objectively call target genes from ChIP-chip data (**Chapter 12**). In a second approach, DNAse hypersensitivity chip (**Chapter 13**), protein-associated regulatory regions are inferred by their chromatin accessibility and selective sensitivity to DNAse digestion, identified by labeling cut DNA ends followed by hybridization to promoter/tiling arrays.

5. DNA Replication

DNA replication is a fundamental process of all cells. Yet, in higher eukaryotes little has been known of the global orchestration and timing of DNA replication, and whether these events might differ in diseases like cancer. In this volume, a method is presented to map the origins and progression of DNA replication, by isolating newly replicated DNA, and then labeling and hybridizing to genome-tiling DNA microarrays (**Chapter 14**).

6. Integrating Data Across Platforms

Characterizing DNA CNAs, methylation patterns, DNA–protein interactions, and resultant transcriptional profiles each provide important information on biological states in health and disease. However, integrating these diverse data types can provide even greater insight. For example, mapping DNA amplifications can pinpoint cancer genes, but does not distinguish "driver" onco-genes from co-amplified neighbors. Likewise, transcriptional pro-filing does not distinguish overexpressed oncogenes from other genes overexpressed secondary to oncogenesis, reflecting for example the proliferative or differentiation state of cancer cells. Yet intersecting amplified and overexpressed genes can effectively enrich driver cancer genes. In this volume, a statistical method, "DR.-Integrator", is presented for integrating genomic and transcriptional profiles for cancer gene discovery (**Chapter 15**).

7. Looking Forward

Over the past few years, DNA microarrays have provided a powerful approach to characterize the physical genome. In the past 1–2 years, ultra-high-throughput DNA sequencing (UHTS) methods have begun to furnish an alternative approach. Commercial platforms like Roche/454, Illumina/Solexa, and ABI utilize cyclic array sequencing to delineate short sequence reads in a massively parallel manner over hundreds of thousands of arrayed DNA sequence features. Counting sequence tags can provide a quantitative assessment of transcripts, genomic DNA copy, methylated DNA, and sites of protein–DNA interaction *(10)*.

Nonetheless, hybridization-based microarray approaches, as detailed in this volume, continue to provide advantages in being tried-and-true, lower cost, and accessible technologies (both in regards to instrumentation and informatics infrastructure). Further, the key "front end" sample preparation steps of the protocols provided here are typically shared in both hybridization and sequencing-based microarray approaches. This compilation of methods and analysis/informational chapters should provide a valuable guide and reference for the molecular or computational biologist interested in characterizing the physical genome, in particular in relation to cancer.

References

1. Fodor, S. P., Rava, R. P., Huang, X. C., Pease, A. C., Holmes, C. P., and Adams, C. L. (1993) Multiplexed biochemical assays with biological chips. Nature **364**, 555–556.

2. Schena, M., Shalon, D., Davis, R. W., and Brown, P. O. (1995) Quantitative monitoring of gene expression patterns with a complementary DNA microarray. Science **270**, 467–470.

3. Alizadeh, A. A., Eisen, M. B., Davis, R. E., Ma, C., Lossos, I. S., Rosenwald, A. et al. (2000) Distinct types of diffuse large B-cell lymphoma identified by gene expression profiling. Nature **403**, 503–511.

4. Perou, C. M., Sorlie, T., Eisen, M. B., van de Rijn, M., Jeffrey, S. S., Rees, C. A., et al. (2000) Molecular portraits of human breast tumours. Nature **406**, 747–752.

5. van't Veer, L. J., Dai, H., van de Vijver, M. J., He, Y. D., Hart, A. A., Mao, M., et al. (2002) Gene expression profiling predicts clinical outcome of breast cancer. Nature **415**, 530–536.

6. Lengauer, C., Kinzler, K. W., and Vogelstein, B. (1998) Genetic instabilities in human cancers. Nature **396**, 643–649.

7. Shaffer, L. G., and Bejjani, B. A. (2006) Medical applications of array CGH and the transformation of clinical cytogenetics. Cytogenet Genome Res **115**, 303–309.

8. Pinkel, D., and Albertson, D. G. (2005) Comparative genomic hybridization. Annu Rev Genomics Hum Genet **6**, 331–354.

9. Herman, J. G., and Baylin, S. B. (2003) Gene silencing in cancer in association with promoter hypermethylation. N Engl J Med **349**, 2042–2054.

10. Wold, B., and Myers, R. M. (2008) Sequence census methods for functional genomics. Nat Methods **5**, 19–21.

Chapter 2

Comparative Genomic Hybridization on BAC Arrays

Bradley P. Coe, William W. Lockwood, Raj Chari, and Wan L. Lam

Abstract

Alterations in genomic DNA are a key feature of many constitutional disorders and cancer. The discovery of the underlying regions of gene dosage has thus been essential in dissecting complex disease phenotypes and identifying targets for therapeutic intervention and diagnostic testing. The development of array comparative genomic hybridization (aCGH) using bacterial artificial chromosomes (BACs) as hybridization targets has facilitated the discovery and fine mapping of novel genomic alterations allowing rapid identification of target genes.

In BAC aCGH, DNA samples are first labeled with fluorescent dyes through a random priming reaction with 100–400 ng of genomic DNA. This probe is then co-hybridized to an array consisting of BAC clones, either tiling the genome (~50 kbp resolution) or spaced at intervals (e.g., 1 Mbp resolution). The resulting arrays are then imaged and the signal at each locus is compared between a reference and test sample to determine the copy number status. The DNA samples to be analyzed may be derived from either fresh, frozen, or formalin-fixed paraffin-embedded material, and sample requirements are currently significantly lower than those for oligonucleotide platforms due to the high probe-binding capacity of BAC clone targets (~150 kbp) compared to oligonucleotides (25–80 bp). In this chapter, we describe in detail the technical procedure required to perform copy number analysis of genomes with BAC aCGH.

Key words: CGH, array CGH, bacterial artificial chromosomes, genomics, gene dosage, DNA copy number.

1. Introduction

DNA copy number changes are hallmarks of constitutional diseases and cancer. Somatic alterations in gene dosage lead to the disruption of both oncogene and tumor suppressor gene expression during cancer development whereas variations in DNA copy number have been associated with developmental disorders (1). Therefore, great effort has been employed to define regions of

Jonathan R. Pollack (ed.), *Microarray Analysis of the Physical Genome: Methods and Protocols*, vol. 556
© Humana Press, a part of Springer Science+Business Media, LLC 2009
DOI 10.1007/978-1-60327-192-9_2 Springerprotocols.com

copy number change in these diseases in order to uncover pathologically related genes. The advent of conventional comparative genomic hybridization (CGH) allowed researchers to understand the patterns of gene dosage across the entire genome, albeit at a relatively low resolution of ~10 Mbp (1, 2). This technique can identify regions of DNA duplication, amplification, and deletion in diseased cell populations, but will not detect balanced chromosomal alterations such as translocations (3). The capabilities of chromosome-based CGH were improved by the development of array CGH (aCGH), whereby DNA targets are spotted onto a glass surface to serve as hybridization targets as an alternative to using metaphase chromosome spreads (4, 5). Numerous genome-wide CGH arrays have since been produced each differing in the size of the genomic elements spotted and their corresponding coverage of the human genome (6). The first reported genome-wide aCGH approach was using expression cDNA microarrays (7). The utilization of large insert clones (typically BAC clones of ~100–150 kbp size) improved the sensitivity of hybridization targets to their corresponding probes, resulting in high signal-to-noise ratios and accurate assessment of copy number (8). Most recently, oligonucleotide (25–80 bp nucleotide probes) arrays were also developed with the goal of improving the maximal resolution of CGH beyond the size of a BAC clone. Each technology allows the high-resolution profiling of sample genomes with distinct advantages and disadvantages for each. For example, BAC arrays require far less sample input than oligonucleotide arrays allowing the analysis of low-yield microdissected specimens while oligonucleotides have the potential to offer greater resolving power (1). Given the existence of the various types of array technology used in CGH studies, the types of situations that arrays may be utilized are numerous. Although the most frequent use of aCGH is in the detection of somatic changes pertaining to cancer, they are also widely used in delineating alterations in developmental disorders and aiding evolutionary comparisons.

BAC aCGH hybridization is performed by first generating labeled DNA samples by a random priming-based reaction where cyanine dyes are incorporated into copies of the original DNA template. Due to the linear nature of random prime labeling, probe is generated proportionately to the original DNA copy number for each region. By co-hybridizing samples representing a normal (diploid) and sample specimen, mixed with Cot-1 DNA (to block repetitive elements) to the array, we can then infer copy number from the ratios with which each probe binds to a particular segment of the genome (**Fig. 2.1**). The following protocol is adapted from Ishkanian et al. (8).

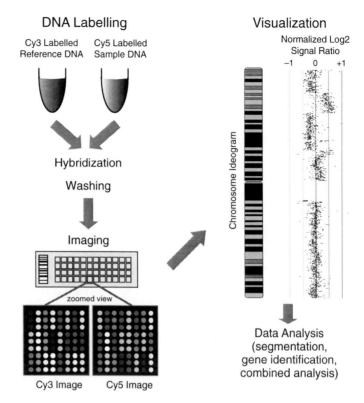

Fig. 2.1. Overview of the array CGH protocol.

2. Materials

2.1. Labeling of Genomic DNA

1. Labeling buffer: Random 8-mer oligonucleotides at 7 µg/µl (Alpha DNA) and Klenow DNA polymerase buffer diluted to 5X concentration (Promega)

2. Klenow DNA polymerase 9 units/µl (Promega), keep on ice at all times

3. Cyanine-3 dCTP 1 nmol/µl (Amersham, GE Healthcare) light-sensitive, pre-aliquot to reduce freeze–thaw cycles

4. Cyanine-5 dCTP 1 nmol/µl (Amersham, GE Healthcare) light-sensitive, pre-aliquot to reduce freeze–thaw cycles

5. RP dNTP mix (2 mM each of dATP, dGTP, dTTP, and 1.2 mM dCTP) (Promega)

6. dH₂O

7. Incubator set at 37°C

2.2. Preparation of Probe

1. Microcon YM-30 filter columns (Millipore)

2. Cot-1 DNA (Invitrogen)

3. DIG Easy hybridization granules, reconstituted at 4.67 g/10 ml (Roche)

4. For option 2: 3 M sodium acetate and 100% ethanol

2.3. Blocking of Repetitive Elements

1. Dry bath incubator set to 85°C
2. Incubator set at 45°C

2.4. Hybridization

1. BAC CGH array (various suppliers)
2. Coverslip 22 × 60 mm (or appropriate size for array used) (Fisher Scientific)
3. Hybridization cassette (Telechem)
4. dH$_2$O
5. Hybridization incubator set to 45°C
6. Optional pre-hybridization buffer: 4.5 µl 20 mg/ml sheared Herring sperm (SHS) DNA, 4.5 µl 10% BSA, 36 µl DIG Easy
7. For optional pre-hybridization: 100% ethanol, isopropanol, and Coplin jar or slide staining dish

2.5. Washing

1. Wash solution 1: 0.1X SSC 0.1% SDS pH 7.0
2. Wash solution 2: 0.1X SSC pH 7.0
3. Coplin jar or slide staining dish
4. 50-ml conical tubes and Eppendorf 5810R centrifuge (or similar) with swinging bucket rotor, or dedicated microarray centrifuge.
5. Dark slide box (for storing arrays)

2.6. Imaging

1. Microarray scanner. Common systems include the Applied Precision Arrayworx, Axon GenePix 4000B or 4200A, and Perkin Elmer ScanArray.

2.7. Analysis

Most array analysis applications are based on the Microsoft Windows© platform and require a standard modern PC. Due to the significant computational load involved in high-resolution data analysis, at least 2–4 GB of RAM is highly recommended.

3. Methods

3.1. Labeling of Genomic DNA

1. Each CGH array will require 100–400 ng of DNA for both a reference and test sample (*see* **Note 1**). DNA samples should be prepared at a concentration of at least 10 ng/µl or higher.
2. For each array prepare two labeling reactions (one for a reference sample and one for the sample you wish to profile).

Combine 100–400 ng DNA (use an equal amount for both the sample and reference) with 5 μl labeling buffer in a 0.2-ml PCR tube. Bring the total reaction volume up to 16.75 μl and boil for 10 min at 100°C.

3. After boiling place the tubes on ice and add the following: 3.75 μl dNTP mix, 2.5 μl Klenow polymerase, and 2 μl of 1 mM Cy3-dCTP or Cy-5 dCTP (*see* **Note 2**). Mix well by pipetting the solution up and down several times.

4. Incubate the labeling reactions overnight at 37°C.

3.2. Preparation of Probe

3.2.1. Option 1: Rapid Protocol

1. Combine the labeling reaction pairs (one reference and one sample) with 100 μl Cot-1 DNA and add to a Microcon YM-30 column.

2. Spin columns at 13,500g for 10 min in the provided tubes.

3. At this point, visual inspection of the DNA pellet in the Microcon can be used to determine if the labeling reaction was successful. A reaction in which both probes have labeled well should appear purple in color, whereas a significant shift to blue or pink represents a failure in one of the labeling reactions and hybridization is likely to be unsuccessful (*see* **Note 3**).

4. Discard the eluate and add 200 μl dH$_2$O to the column

5. Spin columns at 13,500g for an additional 10 min in the provided tubes.

6. Add 45 μl of DIG Easy hybridization buffer (*see* **Note 4**) to each Microcon and allow the probe to resuspend for 10 min at room temperature (RT). Following incubation at RT flip the Microcon into a new tube (provided with the Microcon) and spin at 3,000g for 5 min. (DIG Easy volume is for 22 × 60 mm coverslip, adjust as appropriate for smaller or larger arrays.)

3.2.2. Option 2: Protocol with Labeling Efficiency Analysis

1. Combine the labeling reactions and add to a Microcon YM-30 column.

2. Spin columns at 13,500g for 10 min in the provided tubes.

3. Add 50 μl of dH$_2$O to each column and allow the probe to resuspend for 10 min at RT. Following incubation at RT flip the Microcon into a new tube (provided with the column) and spin at 3,000g for 5 min.

4. At this point the probe may be inspected for the efficacy of the labeling reaction. A reaction in which both probes have labeled well should appear purple in color; this is easiest to observe prior to adding water in Step 3. If a NanoDrop

spectrometer is available it can be used to perform a detailed analysis of dye incorporation, with a 1.5-µl aliquot of the labeled material (*see* **Note 5**).

5. Combine the cleaned up probes with 100 µl Cot-1 DNA and precipitate by adding 15 µl of 3 M NaOAc and 375 µl of 100% EtOH.

6. Incubate at –20°C for ~30 min.

7. Centrifuge at maximum speed in a microcentrifuge for 10 min at 4°C.

8. Remove the supernatant and air dry the pellet for a few minutes so that no ethanol remains.

9. Resuspend the precipitated pellet in 45 µl of DIG Easy hybridization buffer (*see* **Note 4**). (Pre-warming the DIG Easy buffer to 45°C speeds up resuspension; DIG Easy volume is for 22 × 60 mm coverslip, adjust as appropriate for smaller or larger arrays.)

3.3. Blocking of Repetitive Elements with CoT-1 DNA

1. Denature the DIG Easy Probe Solution at 85°C for 10 min.

2. Incubate the DIG Easy Probe Solution at 45°C for 1 h to block repetitive elements (Cot-1 DNA was added in **Section 3.2, Step 1**). Avoid incubating for longer than 1 h to reduce probe self-hybridization.

3.4. Hybridization

3.4.1. Optional: Pre-hybridization Only for BAC Arrays Printed on Amine Slides

Depending on the array being used a pre-hybridization step may be required to block the unreacted sites on the slide used to bind the spotted DNA. In the case of aldehyde-coated slides this is not necessary as a chemical inactivation process is performed shortly after array spotting. For amine-coated slides a pre-hybridization step is required (*see* **Note 6**).

1. Place array in boiling water for 15 s to denature spotted DNA. Then dip slide in ice-cold ethanol to prevent renaturing.

2. Allow slide to air dry by placing on an angle leaning against a pipette tip rack sitting on lint-free wipes. Placing the barcode end toward the bottom will prevent the accumulation of any residue on the hybridization area.

3. Pipette 45 µl of pre-hybridization solution onto the surface of a microarray. Gently lower a 22 × 60 mm coverslip over the probe solution avoiding bubbles, some users may find it easier to place the probe solution on the coverslip and lower the slide onto it. (At this point it is important to proceed through Steps 4 and 5 rapidly to avoid evaporation of probe solution.)

4. Place the array in a hybridization cassette (Telechem) and add 15 µl of water to the lower groove (for humidity control).

5. Seal the hybridization cassette and transfer to a 45°C incubator for 1 h.

6. Remove coverslip and dip the slide in isopropanol. Allow the slide to air dry prior to proceeding to hybridization of probe

3.4.2. Hybridization

1. Pre-heat the microarray slide (pre-hybridized according to **Section 3.4.1** if the array is printed on an amine slide) to 45°C on a slide warmer. Although not strictly necessary, this step helps reduce background hybridization which can occur where the probe is deposited if the coverslip is not applied rapidly.

2. Pipette 45 μl of the probe onto the surface of a microarray. Gently lower a 22 × 60 mm coverslip over the probe solution avoiding bubbles; some users may find it easier to place the probe solution on the coverslip and lower the slide onto it. (At this point it is important to proceed through Steps 3 and 4 rapidly to avoid evaporation of probe solution.)

3. Place the array in a hybridization cassette (Telechem) and add 15 μl of water to the lower groove (for humidity control).

4. Seal the hybridization cassette and transfer to a 45°C incubator for 36–40 h.

3.5. Washing

1. Remove slides from cassettes. Place in pre-warmed wash solution for 1 min to loosen coverslip.

2. If coverslip does not fall off in initial wash, remove coverslip by gently sliding partway off of the slide and then gripping the exposed edge to lift off.

3. Wash slides five times for 5 min at 45°C (agitating).

4. Rinse slides three times to remove residual SDS prior to scanning.

5. Dry slides by centrifugation at 800*g* for 5 min in 50 ml conical tubes, or by using an oil-free air stream, or a dedicated microarray centrifuge. It is important that slides are dried immediately after washing due to potential degradation of cyanine dyes by environmental factors (*see* **Note 7**).

6. Store slides and prepare for scanning.

3.6. Imaging

1. After washing, the microarray may be scanned using any commercial microarray scanner. Important factors to consider include scan resolution and scan intensity (*see* **Note 8**).

2. Most scanners offer various output formats; the most universal is to save a single 16-bit tiff image per cyanine dye. Multiple image tiffs are more convenient when using software such as GenePix but may need to be converted for use with third party applications.

3. Softworx and GenePix are the best automated spot-finding applications in our experience. If your software package

allows automated spot-finding after scanning, it is recommended to visually confirm the quality of the spot-finding especially for arrays with printing artifacts as they may often offset grid placement which can result in erroneous data. This is particularly important for spotted arrays which tend to demonstrate more variable spot placement than on slide-synthesized platforms such as most modern oligonucleotide platforms. Following spot-finding, data should be exported. The most common output formats are tab-delimited text files or GPR files (GPR files are GenePix Results format which is a delimited text file with extra information in the header.)

3.7. Analysis

1. With an output file from imaging analysis, the next necessary step prior to delineating gains and losses is normalization. For BAC arrays, any two-channel normalization approach can be applied to remove systematic biases which may be present during the hybridization experiment. Such factors which need to be accounted for include slide gradients and other intensity-related biases. CGH-Norm *(9)* and MANOR *(10)* are two example programs that can be used for normalization.

2. After the data have been normalized, the process of identifying gains and losses, and the genes encompassed, can be performed. There are two necessary components to do this: interactive visualization and statistical analysis. Interactive visualization is important as it provides genomic context to the identified gains and losses. Genomic context includes, but is not limited to, location of transcribed genes and copy number polymorphisms (changes in copy number found in the normal population). Statistical analysis of aCGH data primarily involves segmentation analysis which is a process whereby breakpoints, points at which a change in copy number occurs, are determined for a given sample. There a number of software packages that are freely or commercially available to aid in this process with some software packages providing either interactive visualization or statistical analysis, while others provide both (**Table 2.1**).

4. Notes

1. This protocol has been validated on formalin-fixed, paraffin-embedded samples and high-quality materials. We have observed that results are dependent on DNA quality. There are many factors which affect the usefulness of an archival sample including age of the block and how the tissue was

Table 2.1
Software for the visualization and/or analysis for aCGH

Software	Cost	Interactive visualization	Statistical analysis	Website
aCGH-Smooth	Free		✓	www.few.vu.nl/~vumarray/
BlueFuse for Microarrays	Cost	✓	✓	www.cytochip.com
CGHcall	Free		✓	www.few.vu.nl/~mavdwiel/ CGHcall.html
CGHFusion	Cost	✓	✓	www.infoquant.com/index/ cghfusion
CGH-Explorer	Free	✓	✓	www.ifi.uio.no/bioinf/Papers/ CGH/
CGH Analytics v3.4	Cost	✓	✓	www.chem.agilent.com/Scripts/ PDS.asp?lPage=29457
DNACopy	Free		✓	bioconductor.org/packages/2.1/ bioc/html/DNAcopy.html
GLAD	Free		✓	bioconductor.org/packages/2.1/ bioc/html/GLAD.html
ISACGH	Free	✓	✓	http://isacgh.bioinfo.cipf.es
LSP-HMM	Free		✓	http://www.cs.ubc.ca/~sshah/acgh
M-CGH	Free	✓	✓	folk.uio.no/junbaiw/mcgh/
MD-SeeGH	Free	✓	✓	www.flintbox.com/ technology.asp?page=706
Nexus CGH	Cost	✓	✓	http://www.biodiscovery.com/ index/nexus
SIGMA^2	Free	✓	✓	sigma.bccrc.ca/sigma2
SpectralWare	Cost	✓	✓	las.perkinelmer.com/Catalog/ ProductInfoPage.htm?Product ID=5007-1010
STAC	Free		✓	cbil.upenn.edu/STAC/
VAMP	Free	✓	✓	bioinfo.curie.fr/vamp

stored prior to embedding (i.e., time in formamide) *(11)*. DNA size is a good starting point to determine DNA quality and we recommend DNA with an average size above 300 bp; however, more degraded samples may work and, depending on sample rarity, they may be worth examining. Care must be taken in examining data from archival material as degraded

DNA can generate false copy number alterations *(11)*. The false copy number profile closely follows the GC% of the genome so corrections may also be applied using this information. Additionally, clustering analysis is an excellent way to identify the data demonstrating the false pattern as it is well preserved across different samples.

Reference samples can be a pooled reference from a commercial source (Novagen), a single common reference, or a matched reference sample. We commonly hybridize tumor samples and matched normal samples separately to prevent masking of non-somatic alterations, such as CNVs or cancer-predisposing regions. An example of the value of this can be seen in Coe et al. *(12)* where we identified a copy number alteration in normal cells from a patient with SCLC which contains various apoptosis genes.

2. The choice of which cyanine dye to use for sample and reference material is up to personal choice, so long as a consistent order is used in all experiments to be compared. For cases when the highest quality data is essential, a dye flip reaction can be performed where two aCGH experiments are performed each with the dye used for the sample changed. By combining the two array results, one can generate a result which has reduced noise compared to a single experiment-based result, beneficial in detecting low level or very small alterations with increased confidence *(13, 14)*.

3. In our experience, hybridization quality results can be predicted fairly accurately using simple visual analysis of the probe color. However, we would recommend more stringent quality control provided in probe preparation option 2 (*also see* **Note 5**) in clinical environments or when arrays are more precious than samples.

4. DIG Easy hybridization buffer is a proprietary hybridization solution which demonstrates similar performance to formamide-based solutions. The primary benefit of this solution is the lower toxicity compared to formamide-containing buffers.

5. If dye incorporation is less than 3.0 pmol/μl, poor results are usually observed. Typical random prime labeling yields are approximately 10 μg, which corresponds roughly to 0.5% of bases labeled. In our experience ideal labeling reactions will produce incorporations of 8–25 pmol/μl per dye.

6. The two most common array chemistries are amine and aldehyde. Amine surfaces bind the spotted DNA by charge interactions whereas aldehyde slides bind through a covalent bond formed during slide processing *(15, 16)*. As amine slides retain their reaction chemistry after processing, pre-hybridization is

required to block amine site on the slide from binding the probe resulting in high background signal. The presented pre-hybridization protocol assumes that arrays have been previously processed to crosslink the spotted BACs to the glass slide. If this has not been performed, the DNA should be crosslinked according to the manufacturer's protocol.

7. Cyanine dyes are susceptible to degradation by ozone *(17, 18)*. Cyanine 5 is particularly sensitive and normal atmospheric ozone levels may completely degrade the dye in certain environments (cities, certain times of day, and spring/summer). There are several solutions available to counteract the effects of ozone on cyanine dyes including air filtration systems with carbon filters (various suppliers, www.iqair.us) or ozone catalysts (See The Pat Brown Lab protocols page for details http://cmgm.stanford.edu/pbrown/protocols/index.html) and chemical slide treatments such as Agilent Stabilization and Drying solution, and addition of antioxidants such as cysteamine to wash buffers *(19)*. In addition to environmental controls, we recommend investing in an ozone sensor to monitor atmospheric conditions in your laboratory, and reducing the exposure time of cyanine dyes to air particularly during the slide drying steps where the dye is most susceptible to degradation. This is especially important to consider when using an array scanner with an autoloader which will hold multiple arrays for scanning.

8. There are two main types of microarray scanner: CCD and laser-based systems. Both are appropriate for the analysis of BAC arrays; however, a few differences should be understood. For laser-based scanners such as the Axon GenePix and PerkinElmer ScanArray systems, a scan resolution of 5 μm or 10 μm, whichever is closest to 10 pixel diameters per spot, is optimal (effects both automated spot-finding by improving feature morphology, and data reliability by increasing the number of measurements used to determine a spot's average intensity). Thus, for an array of 120 μm spots, a 10 μm scan resolution is ideal. Scan intensities may be adjusted manually, or often an automatic adjustment option is available. This is a highly recommended setting as the scanner will automatically adjust the images for optimal dynamic range. For particularly dim images, the option of scanning a slide multiple times and averaging the images to remove noise is worth considering.

For CCD-based systems, resolutions may be offered in increments other than 5 μm, such as the 3.25 μm resolution of the API ArrayWorx. In CCD-based systems, lower resolutions are accomplished by binning pixels together, thus allowing higher intensities to be generated in low-resolution scans (for most purposes a resolution of 9.75 μm is optimal). The binning effect of low resolutions is important to consider when dealing with low-intensity arrays, where reliable signal may require a lower-resolution scan than with a laser system which runs at lower resolutions by simply skipping alternating points on the slide. Be careful to ensure that the back surface of the slide is clean, as many CCD-based systems scan through the back of the slide and are sensitive to out-of-focus fluorescent artifacts on the back of the slide.

Acknowledgments

The authors would like to thank Chad Malloff, Heather Saprunoff, Spencer Watson, and Emilie Vucic for useful discussion. This work was supported by funds from the Canadian Institutes for Health Research, Canadian Breast Cancer Research Alliance, Genome Canada/British Columbia, and National Institute of Dental and Craniofacial Research (NIDCR) grant R01 DE15965.

References

1. Lockwood WW, Chari R, Chi B, Lam WL. (2006) Recent advances in array comparative genomic hybridization technologies and their applications in human genetics. *Eur J Hum Genet.* **14**, 139–148.

2. Kallioniemi A, Kallioniemi OP, Sudar D, Rutovitz D, Gray JW, Waldman F, Pinkel D. (1992) Comparative genomic hybridization for molecular cytogenetic analysis of solid tumors. *Science.* **258**, 818–821.

3. Albertson DG, Collins C, McCormick F, Gray JW. (2003) Chromosome aberrations in solid tumors. *Nat Genet.* **34**, 369–376.

4. Pinkel D, Segraves R, Sudar D, Clark S, Poole I, Kowbel D, Collins C, Kuo WL, Chen C, Zhai Y, Dairkee SH, Ljung BM, Gray JW, Albertson DG. (1998) High resolution analysis of DNA copy number variation using comparative genomic hybridization to microarrays. *Nat Genet.* **20**, 207–211.

5. Solinas-Toldo S, Lampel S, Stilgenbauer S, Nickolenko J, Benner A, Dohner H, Cremer T, Lichter P. (1997) Matrix-based comparative genomic hybridization: biochips to screen for genomic imbalances. *Genes Chromosomes Cancer.* **20**, 399–407.

6. Davies JJ, Wilson IM, Lam WL. (2005) Array CGH technologies and their applications to cancer genomes. *Chromosome Res.* **13**, 237–248.

7. Pollack JR, Perou CM, Alizadeh AA, Eisen MB, Pergamenschikov A, Williams CF, Jeffrey SS, Botstein D, Brown PO. (1999) Genome-wide analysis of DNA copy-number changes using cDNA microarrays. *Nat Genet.* **23**, 41–46.

8. Ishkanian AS, Malloff CA, Watson SK, DeLeeuw RJ, Chi B, Coe BP, Snijders A, Albertson DG, Pinkel D, Marra MA, Ling V, MacAulay C, Lam WL. (2004) A tiling resolution DNA microarray with complete coverage of the human genome. *Nat Genet.* **36**, 299–303.

9. Khojasteh M, Lam WL, Ward RK, MacAulay C. (2005) A stepwise framework for the normalization of array CGH data. *BMC Bioinformatics.* **6**, 274.

10. Neuvial P, Hupe P, Brito I, Liva S, Manie E, Brennetot C, Radvanyi F, Aurias A, Barillot E. (2006) Spatial normalization of array-CGH data. *BMC Bioinformatics.* **7**, 264.

11. Mc Sherry EA, Mc Goldrick A, Kay EW, Hopkins AM, Gallagher WM, Dervan PA. (2007) Formalin-fixed paraffin-embedded clinical tissues show spurious copy number changes in array-CGH profiles. *Clin Genet.* **72**, 441–447.

12. Coe BP, Lee EH, Chi B, Girard L, Minna JD, Gazdar AF, Lam S, MacAulay C, Lam WL. (2006) Gain of a region on 7p22.3, containing MAD1L1, is the most frequent event in small-cell lung cancer cell lines. *Genes Chromosomes Cancer.* **45**, 11–19.

13. Dabney AR, Storey JD. (2007) A new approach to intensity-dependent normalization of two-channel microarrays. *Biostatistics.* **8**, 128–139.

14. Dobbin KK, Kawasaki ES, Petersen DW, Simon RM. (2005) Characterizing dye bias in microarray experiments. *Bioinformatics.* **21**, 2430–2437.

15. Taylor S, Smith S, Windle B, Guiseppi-Elie A. (2003) Impact of surface chemistry and blocking strategies on DNA microarrays. *Nucleic Acids Res.* **31**, e87.

16. Zammatteo N, Jeanmart L, Hamels S, Courtois S, Louette P, Hevesi L, Remacle J. (2000) Comparison between different strategies of covalent attachment of DNA to glass surfaces to build DNA microarrays. *Anal Biochem.* **280**, 143–150.

17. Branham WS, Melvin CD, Han T, Desai VG, Moland .CL, Scully AT, Fuscoe JC. (2007) Elimination of laboratory ozone leads to a dramatic improvement in the reproducibility of microarray gene expression measurements. *BMC Biotechnol.* **7**, 8.

18. Fare TL, Coffey EM, Dai H, He YD, Kessler DA, Kilian KA, Koch JE, LeProust E, Marton MJ, Meyer MR, Stoughton RB, Tokiwa GY, Wang Y. (2003) Effects of atmospheric ozone on microarray data quality. *Anal Chem.* **75**, 4672–4675.

19. Fiegler H, Redon R, Carter NP. (2007) Construction and use of spotted large-insert clone DNA microarrays for the detection of genomic copy number changes. *Nat Protoc.* **2**, 577–587.

Chapter 3

Comparative Genomic Hybridization on Spotted Oligonucleotide Microarrays

Young H. Kim and Jonathan R. Pollack

Abstract

Recent advances in DNA microarray technology have enabled researchers to comprehensively characterize the complex genomes of higher eukaryotic organisms at an unprecedented level of detail. Array-based comparative genomic hybridization (Array-CGH) has been widely used for detecting DNA copy number alterations on a genomic scale, where the mapping resolution is limited only by the number of probes on the DNA microarray. In this chapter, we present a validated protocol utilizing print-tip spotted HEEBO (Human Exonic Evidence Based Oligonucleotide) microarrays for conducting array-CGH using as little as 25 ng of genomic DNA from a wide variety of sources, including cultured cell lines and clinical specimens, with high spatial resolution and array-to-array reproducibility.

Key words: DNA microarray, array-CGH, comparative genomic hybridization, HEEBO, post-processing, epoxysilane, whole-genome amplification.

1. Introduction

Oligonucleotide-based microarrays consist of thousands of defined oligonucleotide hybridization probes ranging from 25 to 85 nucleotides in length, that are either robotically spotted or synthesized in situ in a pattern of rows and columns on a microscope slide or a chip (1). Oligo-based microarrays have been extensively utilized for genome-wide expression profiling studies in a number of eukaryotic organisms (2–7). However, the recent availability of the complete genome sequences for many of these organisms has extended the utility of oligonucleotide arrays to genomic analysis, including the characterization of genomic

Jonathan R. Pollack (ed.), *Microarray Analysis of the Physical Genome: Methods and Protocols, vol. 556*
© Humana Press, a part of Springer Science+Business Media, LLC 2009
DOI 10.1007/978-1-60327-192-9_3 Springerprotocols.com

DNA copy number aberrations such as gene amplifications and deletions in human cancers *(8–12)* and high-resolution studies of human genetic disorders *(13–15)*.

Array-based oligonucleotide comparative genomic hybridization (array-CGH) has rapidly gained acceptance as a technique of choice to characterize DNA copy number alterations at a high resolution. To measure gene copy number, genomic DNAs from two different samples (e.g., tumor and normal) are differentially labeled with fluorescent dyes and co-hybridized to an oligonucleotide microarray that is then scanned, and fluorescent intensities for each oligo element on the array reported. The ratio of fluorescence intensities for each oligo element on the array reflects the relative abundance of the corresponding gene's (or locus') copy number between the two samples (**Fig. 3.1**).

There are several potential advantages of oligo arrays over cDNA and bacterial artificial chromosome (BAC) platforms for genomic analysis. In theory, oligo arrays can be designed for any organism whose genomic sequence is known, while coverage of cDNA and BAC microarrays is limited to the availability of physical clones in dbEST and genomic clone libraries, respectively. Oligo arrays also circumvent some of the technical issues associated with library-based platforms including clone generation, tracking and validation, and offer flexibility in design and synthesis resulting in significant time and cost savings *(10)*. Finally, like cDNA-based arrays, the same oligo microarray may be designed and used for both array-CGH and gene expression analysis which facilitates the comparison of DNA copy number and gene expression data *(16)* (**Fig. 3.2**).

In this chapter, we present details requisite to successful application of the array-CGH technique on spotted "home-made" HEEBO microarrays, including *(1)* preparation, labeling, and hybridization of genomic DNA, *(2)* processing of microarrays, and *(3)* post-hybridization protocols, including microarray washing, imaging, and data analysis. While described for the HEEBO array platform, this is a generic protocol that should be equally applicable to any other spotted oligonucleotide microarray platform.

2. Materials

2.1. HEEBO Microarrays

2.1.1. About HEEBO Microarrays

High-quality spotted HEEBO 70-mer microarrays covering a majority of human genes and providing an average genome mapping resolution of ~35 kb are readily available from various academic microarray core facilities (e.g., Stanford Functional Genomics Facility) or commercial sources for immediate use in microarray

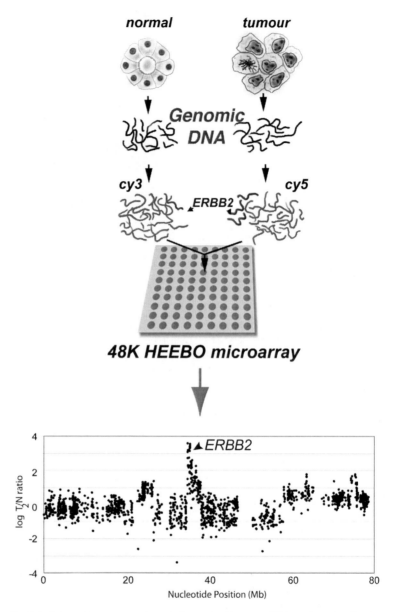

Fig. 3.1. Schematic representation of oligonucleotide array-CGH and data visualization. Test (tumor) and reference (normal) samples are differentially labeled with Cy5/Cy3 fluorophores and co-hybridized to a HEEBO microarray containing ∼48,000 oligonucleotide probe features. Arrays are scanned and the ratio of fluorescence intensities for each array element indicates the relative gene copy number (i.e., amplification or deletion) between the test and reference samples. Fluorescence ratios for each probe are ordered by genomic position using a genome reference assembly (e.g., UCSC GoldenPath). Here, the *ERBB2* gene is shown to be amplified in the tumor sample (indicated by *arrow*).

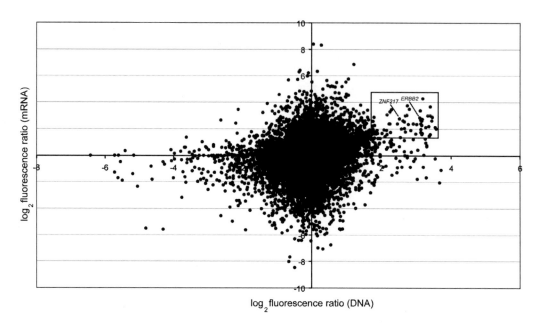

Fig. 3.2. Parallel analysis of DNA copy number and mRNA expression in breast cancer cell line SKBR3 using HEEBO microarrays. Log₂ fluorescence ratios for DNA and mRNA measurements for each gene are plotted. Genes with high-level DNA copy number and mRNA expression values are *highlighted*. Known breast cancer oncogenes (*ERBB2*, *ZNF217*), whose expression is driven by underlying DNA amplification, are indicated.

experiments. Complete details regarding HEEBO array design and feature annotations are available at http://www.microarray.org/sfgf/heebo.do (*see* **Note 1**).

2.1.2. Processing of HEEBO Arrays

1. 2X SSC.
2. 0.1% Triton X-100.
3. 1 mM HCl.
4. 100 mM KCl.
5. 1X Blocking Solution: 50 mM ethanolamine, 0.1% SDS in 0.1 M TrisHCl, pH 9.0.
6. Humidifying slide chambers (Sigma-Aldrich).
7. Metal slide rack and glass staining dishes (Wheaton Science).
8. Temperature-controlled shaking water bath (Bellco Glass).
9. Desktop centrifuge with microtiter plate adaptors (Beckman Allegra GS-6R or equivalent).
10. Slide warmer (LabScientific).

2.2. Amplification and Labeling of Genomic DNA

1. Illustra GenomiPhi V2 DNA Amplification Kit (GE Healthcare).
2. *Dpn*II restriction enzyme and supplied 10X restriction enzyme buffer (New England Biolabs).

3. BioPrime Array CGH Genomic Labeling Module (Invitrogen): contains 10X dCTP nucleotide labeling mix, high-concentration exo-minus Klenow DNA polymerase (40 U/μL), 2.5X random primers (octamers), and stop solution (0.5 M EDTA).

4. Cy5-dCTP, Cy3-dCTP (GE Healthcare).

2.3. DNA Clean-Up and Hybridization

1. TE pH 7.4.

2. Microcon YM-30 filters (42410; Fisher Scientific).

3. Human Cot-1 DNA (1 μg/μL) (15279-011; Invitrogen).

4. Oligo aCGH/ChIP-on-ChIP Hybridization Kit (contains 2X Hi-RPM Hybridization Buffer and 10X Lyophilized Blocking Agent) (Agilent Technologies).

5. SureHyb Hybridization Chambers (Agilent Technologies).

6. Hybridization Slide Gaskets (Agilent Technologies).

7. Microarray Hybridization Oven (Agilent Technologies).

8. Rotator Rack for Hybridization Oven (Agilent Technologies).

2.4. Post-hybridization Washing of Microarrays

1. Wash solution 1: 2X SSC/0.03% SDS.

2. Wash solution 2: 2X SSC.

3. Wash solution 3: 1X SSC.

4. Wash solution 4: 0.2X SSC.

5. Temperature-controlled shaking water bath (Bellco Glass).

6. Metal slide racks *(2)* and glass staining dishes (Wheaton Science).

7. Desktop centrifuge with microtiter plate adaptors (Beckman Allegra GS-6R or equivalent).

2.5. Microarray Imaging, Data Acquisition and Analysis

1. Dual-laser confocal microarray scanner (GenePix 4000B; Molecular Devices or equivalent) for microarray image acquisition.

2. Feature extraction software (SpotReader 1.3; Niles Scientific or equivalent) for automatic identification and extraction of fluorescent ratios from array elements.

3. Database software (e.g., Microsoft Excel or other customized server-based microarray databases) for data storage, manipulation, and analysis.

3. Methods

3.1. Processing of HEEBO Microarrays

Microarrays will need to be "post-processed" (i.e., processed after printing) to be activated for use in hybridization. Post-processing entails *(1)* hydration of microarrays to enable covalent immobilization of oligonucleotides to reactive epoxysilane moieties on slide surface; *(2)* washing of unbound oligos and residue from array printing; and *(3)* blocking of reactive groups in non-printed areas of the microarray.

3.1.1. DNA Immobilization

1. Warm 100 mL of 2X SSC solution to ~50°C in a microwave oven and pour solution in a humidifying chamber.
2. Carefully place microarrays (printed side down) into the slots of the chamber and cover with lid. Hydrate for at least 1 h at room temperature (*see* **Note 2**).
3. Proceed to washing and blocking.

3.1.2. Washing and Blocking of Microarrays

1. Transfer arrays to a metal slide rack and wash in 0.1% Triton X-100 for 5 min on an orbital shaker at room temperature (*see* **Note 3**).
2. Wash arrays 2X in 1 mM HCl solution for 2 min each on an orbital shaker at room temperature.
3. Incubate arrays in 100 mM KCl solution for 10 min on an orbital shaker at room temperature.
4. Wash arrays for 2 min in ddH$_2$O on an orbital shaker at room temperature.
5. Incubate arrays in pre-warmed 1X Blocking Solution for 20 min in a shaking water bath preset to 50°C.
6. Wash arrays in ddH$_2$O for 5 min at room temperature on an orbital shaker.
7. Quickly transfer slide rack containing arrays to a desktop centrifuge with microtiter adaptors and centrifuge at 500*g* for 5 min.
8. Check individual arrays under ambient light to ensure that there are no residues on array surface (*see* **Note 4**).
9. Incubate arrays at 37°C in a slide warmer until ready for use.

3.2. Genomic DNA

3.2.1. Isolation of Genomic DNA

Isolation of high-quality genomic DNA from cultured cell lines and tissue (*see* **Note 5**) is routinely done using commercially available kits standard in many laboratories. Anion-exchange column-based kits (e.g., Blood and Cell Culture Kit; DNeasy/ QiaAmp Tissue Kit; Qiagen) are good alternatives to traditional genomic DNA isolation methods. Isolated genomic DNA should be run out on an agarose gel to assess quality (*see* **Note 6**), and quantified using ultraviolet spectrophotometry. If less than 3 μg of genomic DNA input is available for subsequent labeling, then follow the optional whole-genome amplification (WGA) proto-col below.

3.2.2. Amplification of Genomic DNA

1. In separate microfuge tubes, dilute test and reference genomic DNA in ddH$_2$O to a concentration of 25 ng/μL (*see* **Note 7**).

2. Add 1 μL diluted genomic DNA from test and reference samples into separate microfuge tubes and mix with 9 μL of Sample Buffer (this and other reagents are provided with the Illustra GenomiPhi V2 DNA Amplification Kit).

3. Heat denature samples in a boiling water bath for 3 min, and immediately snap-cool samples on ice.

4. For each amplification reaction, prepare a master mix containing 9 μL Reaction Buffer and 1 μL Enzyme Mix, and add 10 μL to each test and reference DNA sample. Mix well and briefly spin-down in a microcentrifuge (*see* **Note 8**).

5. Incubate samples at 37°C for 1.5 h.

6. Heat-inactivate enzyme at 65°C for 10 min.

7. Quantify amplified DNA yield by UV spectrophotometry and check quality (confirming adequate DNA size) by agarose gel electrophoresis of a small aliquot (*see* **Note 9**).

3.2.3. Labeling of Genomic DNA

1. In separate microfuge tubes, aliquot and digest 3 μg of (amplified) test and reference genomic DNAs using *Dpn*II restriction enzyme as per manufacturer's directions (*see* **Note 10**).

2. Heat-inactivate digestion reaction by incubation at 65°C for 20 min.

3. Add 20 μL 2.5X random primer mix (this and other reagents are provided with the BioPrime Array CGH Genomic Labeling Module) to each tube and place samples in a boiling water bath for 5 min, then snap-cool samples on ice for another 5 min (*see* **Note 11**)

4. To each tube, add 5 μL 10X dCTP nucleotide labeling mix, 3 μL Cy5-dCTP (for test/tumor DNA) or Cy3-dCTP (for reference/normal DNA), and 1 μL high-concentration Klenow enzyme.

5. Incubate samples at 37°C for 2 h.

6. Stop the reactions by adding 5 μL of the stop solution to each tube.

3.3. DNA Clean-Up and Hybridization

1. Pool paired test and reference samples together in a Microcon YM-30 filter with 400 μL TE (pH 7.4) and spin at 12,000g in a microcentrifuge for 10–12 min.

2. Discard the flowthrough and add an additional 500 μL TE (pH 7.4) to each sample, re-centrifuge at 12,000g for 10–12 min. Carefully measure the remaining volume in the filter (should be <20 μL). Discard flowthrough (*see* **Note 12**).

3. Invert filter to a clean microfuge tube and spin at 12,000g for 1 min to collect the labeled probe mixture.

4. For each sample, bring up the volume to 150 μL with ddH$_2$O. Add 50 μL of Human Cot-1 DNA, 50 μL of 10X Agilent Blocking Agent, and 250 μL of 2X Hi-RPM Hybridization Buffer (*see* **Notes 13** and **14**). Mix gently (to avoid bubbles) and boil hybridization mixture in a heated water bath for 3 min.

5. Incubate samples at 37°C for 30 min.

6. Place ~500 μL of hybridization mixture onto the slide gasket (assembled in hybridization chamber as per manufacturer's instructions) and place microarray with printed side facing the hybridization mix to form an "array sandwich".

7. Load hybridization chambers into a rotating oven set at 65°C with a rotation speed of 20 rpm for 30–40 h (*see* **Note 15**).

3.4. Post-hybridization Washing of Microarrays

Washing of microarrays following hybridization serves to remove unbound labeled nucleic acid.

1. In separate glass staining chambers, prepare 400 mL of each of the four wash solutions, and preheat wash solutions 1 and 2 to 65°C (*see* **Note 16**).

2. Place arrays in a slide rack, and gently agitate for 3 min each sequentially in wash solution 1 (at 65°C), 2 (at 65°C), 3 (at room temperature), and 4 (at room temperature).

3. After the final wash, spin dry microarrays using a desktop centrifuge with microtiter plate adaptors at 500g for 5 min.

3.5. Microarray Imaging, Data Acquisition, and Analysis

Hybridized microarrays should be immediately scanned in a dual-color microarray scanner (e.g., GenePix 4000B) or equivalent (*see* **Note 17**). After image acquisition, fluorescence intensity ratios are extracted from each array feature using a feature extraction software (e.g., GenePix Pro 6.0; Molecular Devices, SpotReader v1.3; Niles Scientific; BlueFuse; Cambridge Blue Gnome). To account for differences in sample input and labeling efficiency, feature-extracted raw data are normalized by setting the average log$_2$ ratio for each array to 0. Resulting microarray data can be stored in a number of databases, ranging from Microsoft Excel to more specialized solutions designed for storage, retrieval, and analysis of large microarray datasets (*see* **Note 18**). For visualization and analysis of HEEBO array-CGH data, academic and commercial solutions are available that map fluorescence ratios of each oligonucleotide on the array to its physical genomic location based on genome assemblies such as UCSC GoldenPath (http://genome.ucsc.edu) (*see* **Note 19**).

4. Notes

1. For array-CGH, it is highly recommended that oligo arrays be printed on microscope slides derivatized with epoxysilane functional groups for best performance.

2. Prior to end of hydration, turn on water bath to 50°C and prepare fresh 1X Blocking Solution. Incubate 1X Blocking Solution in water bath until ready for use.

3. All washing solutions should be made fresh and 0.22-μm sterile-filtered prior to use. Microarrays must be completely submerged in solution at all times during the washing and blocking steps; thus prepare enough of each wash/blocking solution (depending on the size of vessel used) to ensure microarrays are not exposed to air for prolonged periods of time. Triton X-100 solution should be completely dissolved before use; if not dissolved, place in a 50°C water bath for 5 min.

4. It is important to remove any traces of blocking residue from the array surface, as it will cause noticeable fluorescent background when arrays are scanned. If hazy or "swirl" residue is visible on the array surface, wash arrays in molecular biology grade 70% ethanol for 5 min and spin-dry. This wash step should eliminate any remaining surface residue on the array.

5. Microdissection techniques (e.g., laser-capture microscopy) can be used to increase the enrichment of certain populations of interest (e.g., tumor cells) from heterogeneous tissue mixtures.

6. Genomic DNA isolated from archival tissue specimens (e.g., ethanol- or formalin-fixed paraffin-embedded samples) can vary in quality depending on the age of tissue and the method of fixation, and may produce variable hybridization results. In our experience, samples where most DNA is <1 kb (by agarose gel electrophoresis) indicates substantial and biased DNA degradation, and produces noisy array-CGH data. It is also recommended that a test hybridization be performed on a representative archival specimen(s) to assess data quality prior to commencement of a large-cohort microarray study.

7. To increase the amount of starting input material for hybridization, a WGA approach is utilized. This technique takes advantage of several unique properties of the Φ29 phage DNA polymerase, including extreme template processivity, multiple-strand displacement capabilities, and high fidelity. Typical yields of amplified genomic DNA range from 4 to

6 μg, starting from 1 to 25 ng of input. To minimize potential amplification biases, similar amounts of test and reference genomic DNAs should be amplified in parallel.

8. The master mix contains all the necessary components (e.g., salts, random primers, dNTPs, DNA polymerase) to initiate the amplification reaction.

9. The size range of pre-amplification and post-amplification DNAs should be compared by agarose gel electrophoresis. Following a successful amplification reaction, the DNA size range should be ~2–10 kb. Smaller sizes indicate degraded input DNA or unsuccessful amplification.

10. The total volume of the digestion reaction should be <25 μL; thus, input DNA may need to be concentrated first by using a SpeedVac or other methods. Other four-base cutting restriction enzymes (e.g., *Alu*I or *Rsa*I) may be used in lieu of *Dpn*II. Restriction digestion of starting genomic DNA serves to increase labeling efficiency.

11. Boiling of sample ensures complete denaturation of DNA, and subsequent snap-cooling of samples on ice allows random primers to bind to complementary single-stranded DNA template.

12. The additional washing step is necessary to remove unincorporated fluorescent nucleotides. If the remaining volume in the filter is >20 μL, centrifuge samples in additional 1 min increments until the volume is <20 μL. Visualization of a purple-colored probe mixture on the filter is indicative of a successful labeling reaction.

13. Human Cot-1 and 10X Blocking Agent are used for blocking repetitive elements in target DNA and non-specific hybridization, respectively. 10X Blocking Agent should be re-suspended to a final volume of 1,250 μL in ddH$_2$O before use.

14. The Agilent hybridization buffer is convenient and in our experience performs comparably or better than other aqueous hybridization solutions. Also, our experience is that high-volume hybridizations carried out with mixing of hybridization solution (as detailed here) yields stronger fluorescence hybridization signal-to-noise compared to protocols using low-volume hybridizations under a glass coverslip.

15. The optimal length of time for efficient hybridization depends on the hybridization kinetics and buffering of the nucleic acid mixture. Hybridization times less than 30 h lead to incomplete nucleic acid binding (e.g., low feature signals) and >40 h lead to high signals but increased background levels.

16. Time can be saved by using preheated ddH$_2$O (by microwaving) to make wash solutions 1 and 2. Heated wash solutions should be placed in a water bath set to 65°C for the duration of washing.

17. Cyanine-based dyes, especially Cy5, are extremely sensitive to atmospheric ozone. Exposure to 5–10 ppb of ozone for short periods (e.g., 30 min) can adversely affect Cy5 signal and data reproducibility. It is recommended that measures be taken to mitigate this effect, such as the use of an ozone eliminator (e.g., Ozone Interceptor; Ozone Solutions, Inc.) or a home-made ozone scrubber (details at http://cmgm.stanford.edu/pbrown/protocols/Ozone_Prevention.pdf).

18. Specialized academic and commercial microarray databases designed for data storage, retrieval, and basic analysis include Stanford Microarray Database (SMD), AMAD (Another MicroArray Database; UCSF), and GenePix Acuity (Molecular Devices).

19. There are a number of applications that can be utilized to visualize and analyze HEEBO array-CGH data. Some of these applications are dedicated for visualization or analys is only, while others offer a comprehensive solution for the end-user. These include Java Caryoscope (for visualizing aCGH data) (http://dahlia.stanford.edu:meme/caryoscope/index.html) and CGH-Miner (automated gain/loss calling and output visualization) (http://www-stat.stanford.edu/~wpll/CGH-Miner). Comprehensive commercial solutions include CGH Analytics (Agilent Technologies), CGH Nexus (BioDiscovery), and Partek Genomics Suite (Partek).

References

1. Ylstra, B., van den Ijssel, P., Carvalho, B., Brakenhoff, R. H., and Meijer, G. A. (2006) BAC to the future! or oligonucleotides: a perspective for micro array comparative genomic hybridization (array CGH). Nucleic Acids Res. **34**, 445–450.

2. Schena, M., Shalon, D., Davis, R. W., and Brown, P. O. (1995) Quantitative monitoring of gene expression patterns with a complementary DNA microarray. Science **270**, 467–470.

3. Wodicka, L., Dong, H., Mittmann, M., Ho, M. H., and Lockhart, D. J. (1997) Genome-wide expression monitoring in *Saccharomyces cerevisiae*. Nat Biotechnol. **15**, 1359–1367.

4. Golub, T. R., Slonim, D. K., Tamayo, P., Huard, C., Gaasenbeek, M., Mesirov, J. P., et al. (1999) Molecular classification of cancer: class discovery and class prediction by gene expression monitoring. Science **286**, 531–537.

5. Hill, A. A., Hunter, C. P., Tsung, B. T., Tucker-Kellogg, G., and Brown, E. L. (2000) Genomic analysis of gene expression in *C. elegans*. Science **290**, 809–812.

6. Hughes, T. R., Mao, M., Jones, A. R., Burchard, J., Marton, M. J., Shannon, K. W., et al. (2001) Expression profiling using microarrays fabricated by an ink-jet oligonucleotide synthesizer. Nat. Biotechnol. **19**, 342–347.

7. Ramaswamy, S., Ross, K. N., Lander, E. S., and Golub, T. R. (2003) A molecular signature of metastasis in primary solid tumors. Nat. Genet. **33**, 49–54.

8. Lucito, R., Healy, J., Alexander, J., Reiner, A., Esposito, D., Chi, M., et al. (2003)

Representational oligonucleotide microarray analysis: a high-resolution method to detect genome copy number variation. Genome Res. **13**, 2291–2305.

9. Barrett, M. T., Scheffer, A., Ben-Dor, A., Sampas, N., Lipson, D., Kincaid, R., et al. (2004) Comparative genomic hybridization using oligonucleotide microarrays and total genomic DNA. Proc. Natl. Acad. Sci. U S A **101**, 17765–17770.

10. Carvalho, B., Ouwerkerk, E., Meijer, G. A., and Ylstra, B. (2004) High resolution microarray comparative genomic hybridisation analysis using spotted oligonucleotides. J. Clin. Pathol. **57**, 644–646.

11. Zhao, X., Li, C., Paez, J. G., Chin, K., Janne, P. A., Chen, T. H., et al. (2004) An integrated view of copy number and allelic alterations in the cancer genome using single nucleotide polymorphism arrays. Cancer Res. **64**, 3060–3071.

12. Kwei, K. A., Kim, Y. H., Girard, L., Kao, J., Pacyna-Gengelbach, M., Salari, K., et al. (2008) Genomic profiling identifies TITF1 as a lineage-specific oncogene amplified in lung cancer. Oncogene **27**, 3635–3640.

13. Friedman, J. M., Baross, A., Delaney, A. D., Ally, A., Arbour, L., Armstrong, L., et al. (2006) Oligonucleotide microarray analysis of genomic imbalance in children with mental retardation. Am. J. Hum. Genet. **79**, 500–513.

14. Ming, J. E., Geiger, E., James, A. C., Ciprero, K. L., Nimmakayalu, M., Zhang, Y., et al. (2006) Rapid detection of submicroscopic chromosomal rearrangements in children with multiple congenital anomalies using high density oligonucleotide arrays. Hum. Mutat. **27**, 467–473.

15. Balciuniene, J., Feng, N., Iyadurai, K., Hirsch, B., Charnas, L., Bill, B. R., et al. (2007) Recurrent 10q22–q23 deletions: a genomic disorder on 10q associated with cognitive and behavioral abnormalities. Am. J. Hum. Genet. **80**, 938–947.

16. Pinkel, D., and Albertson, D. G. (2005) Array comparative genomic hybridization and its applications in cancer. Nat. Genet. **37 Suppl**, S11–S17.

Chapter 4

Comparative Genomic Hybridization by Representational Oligonucleotide Microarray Analysis

Robert Lucito and James Byrnes

Abstract

The central cause to any cancer ultimately lies in the genome and the initial alterations that result in changes in gene expression that are reflected in the phenotype of the cancer cell. The gene expression data are rich in information but the primary lesions responsible for carcinogenesis are obscured due to the complex cascade of expression changes that can occur. The primary lesions can be characterized by the smallest of point mutations to small insertions and deletions (in/dels) to much larger deletions and amplifications (for simplicity all copy number gains will be referred to as amplifications) as well as balanced or unbalanced translocations. In addition to these mutations there are a myriad of epigenetic alterations that affect the cells phenotype. Any gene if important to tumor growth will be altered by mutation or by deletion/amplification eventually, and if a large number of tumor samples is analyzed the majority of these genes will be detected. This chapter describes a variation of comparative genomic hybridization, called Representational oligonucleotide microarray analysis (ROMA), that surveys reduced-complexity representations of tumor genomic DNA to discover deletions and amplifications (and the underlying cancer genes).

Key words: Representational oligonucleotide microarray analysis, ROMA, comparative genomic hybridization, DNA deletion, DNA amplification.

1. Introduction

A number of methods have been developed to identify the regions deleted or amplified in cancer. The majority of techniques developed incorporated, as did ours, the use of custom-designed microarrays *(1–5)*. The genome is large and hybridization of the whole genome, especially to a solid support such as a microarray, is inefficient. To improve efficiency we incorporated the use of genome complexity reducing "representations" *(6, 7)*. Representations are

Jonathan R. Pollack (ed.), *Microarray Analysis of the Physical Genome: Methods and Protocols, vol. 556*
© Humana Press, a part of Springer Science+Business Media, LLC 2009
DOI 10.1007/978-1-60327-192-9_4 Springerprotocols.com

reproducible samplings of the genome produced by cleavage of the genome with a restriction enzyme, ligation of adaptors, and amplification by polymerase chain reaction (PCR) *(6, 7)*. The complexity reduction is a consequence of the PCR amplification. In a mixture of different-sized amplifiable fragments Taq polymerase prefers the smaller sizes, and after 15–20 cycles of PCR the representation is in the range of 100–1,200 bp; the remaining fragments amplify inefficiently. Representations can be prepared from as little as 10–20 ng of input genomic DNA making representational approaches a major advantage when analyzing microdissected tumor samples.

The level of complexity reduction depends on the restriction endonuclease chosen for the digestion. With restriction endonucleases that have a 6-bp recognition sequence, the site being found relatively infrequently, the complexity is roughly 2–3% of the original genome. If a restriction endonuclease with a four-base recognition sequence is chosen the complexity can be anywhere from 15 to 70% depending on the sequence.

GC-rich sequences are found less frequently than AT-rich sequences. Once the complexity is reduced the hybridization can proceed faster. In addition, due to the reduced complexity the sample or representation is easier to label with fluorescent nucleotide. Both of these factors, decreased complexity and increased labeling efficiency, translate to a higher signal-to-noise ratio for features on the array. The representations are then labeled and hybridized to an array designed to match the representation prepared.

2. Materials

2.1. Extraction of DNA from Small Number of Cells

1. TE pH 8.0: 10 mM Tris-HCl, 1 mM EDTA

2. Yeast tRNA (Sigma). Prepare from lyophilized powder by resuspending in diH$_2$O to a concentration of 10 mg/mL, and aliquot (~100 μL) into microcentrifuge tubes and store at –20°C.

3. 10% SDS (Invitrogen).

4. Proteinase K (Sigma); 20 mg/mL dissolved in diH$_2$O.

5. Phenol (Amresco)

6. Phenol:chloroform (Amresco); includes a buffer that must be added at least ~1 h before use.

7. 3 M Sodium Acetate pH 5.2. Prepare by dissolving 408.1 g sodium acetate (+3H$_2$O) (Fisher Scientific) in 600 mL diH$_2$O or MiliQ water and adjusting the pH to 5.2 with

glacial acetic acid (Fisher Scientific). After adjusting the pH, bring the volume to 1 L with diH$_2$O or MiliQ water and aliquot into 125 mL glass bottles and autoclave for 20 min on the liquid cycle.

8. 70% Ethanol. Prepare from 190 proof benzene-free ethanol (Pharmco).

9. UV spectrophotometer (Nanodrop, or equivalent).

2.2. Restriction of DNA

1. Restriction endonuclease *Bgl*II; 10,000 units/mL (New England Biolabs).

2. Phenol:chloroform (*see* **Section 2.1**).

3. Chloroform:isoamyl alcohol (Amresco).

4. 3 M Sodium Acetate pH 5.2 (*see* **Section 2.1**).

5. 100% Ethanol (Pharmco).

6. 70% Ethanol.

7. Yeast tRNA (*see* **Section 2.1**).

2.3. Ligation of Adaptors

1. IBgl24 (TCAGCATCGAGACTGAACGCAGCA) and IBgl12 (GATCTGCTGCGT), 0.20 μmol or higher HPLC purified (Sigma Genosys or equivalent). Resuspend oligonucleotides to 62 pm/μL, roughly 0.50 μg/μL for 24 mer and 0.25 μg/μL for 12 mer, in diH$_2$O (*see* **Note 1**).

2. T4 DNA ligase (100,000 units) (New England Biolabs).

3. Ligase buffer (supplied with the T4 DNA ligase) (*see* **Note 2**).

4. Thermocycler (MJ Research Tetrad thermocylcer or equivalent).

2.4. Representation Amplification and Clean-Up

1. IBgl24 (TCAGCATCGAGACTGAACGCAGCA), 0.20 μmol or higher (Sigma Genosys or equivalent). Resuspend oligonucleotide to 62 pm/μL, roughly 0.50 ug/uL for 24 mer, in diH$_2$O (*see* **Note 3**).

2. Taq polymerase (Qiagen), provided as a Master Mix which includes MgCl$_2$.

3. Isopropanol (ACS grade) (Fisher Scientific).

4. Phenol:chloroform (*see* **Section 2.1**)

5. 3 M Sodium Acetate pH 5.2 (*see* **Section 2.1**)

6. 70% Ethanol (*see* **Section 2.1**)

7. Thermocycler (MJ Research Tetrad thermocylcer or equivalent).

2.5. Representation Labeling

1. Random nonamers (NNNNNNNNN) (SigmaGenosys or equivalent). The synthesis scale should be at least 0.20 μmol. Resuspend nonamers in diH$_2$O to a concentration of 0.50 μg/μL.

2. dNTPs (Invitrogen). Dilute with diH_2O to the following concentrations: adenine, guanine, and thiamine at 1.2 mM and cytosine at 0.6 mM (*see* **Note 4**).

3. High concentration Klenow fragment, 1,000 units at 500,000 units/mL (New England Biolabs).

4. Klenow buffer or buffer 2 (supplied with Klenow).

5. Cy3 dCTP and Cy5 dCTP (GE Healthcare or Perkin Elmer) (*see* **Note 5**).

6. Cot1 DNA (Invitrogen) (*see* **Note 6**).

7. Yeast tRNA, 10 mg/mL dissolved in diH_2O (Invitrogen or Sigma) (*see* **Note 7**).

8. YM-30 Microcon spin column (Millipore).

9. Low TE: 3 mM Tris-HCl pH 7.5, 0.2 mM EDTA. Prepare with 3 mL 1 M Tris-HCl, pH 7.5, 400 μL 0.5 M EDTA, bring volume to 1 L with diH_2O, then filter into 250 mL plastic filter bottle (Corning) (*see* **Note 8**).

2.6. Pre-hybridization and Hybridization

1. Low TE (*see* **Section 2.5**)
2. Formamide (Amresco).
3. 20XSSC (Invitrogen).
4. 10% SDS (Invitrogen).
5. Whatman filters (VWR).
6. Lifter slips (Erie Scientific; can be purchased through VWR).
7. Custom-designed oligonucleotide microarray (Nimblegen) (*see* **Note 9**).
8. Labeled alignment oligonucleotides (Nimblegen) (*see* **Note 10**).
9. Hybridization oven, such as InSlideOut (Boekel).

2.7. Washing and Scanning

1. 20XSSC (Invitrogen).
2. 10% SDS (Invitrogen).
3. Slide holder (VWR).
4. Microarray scanner (Axon GenePix 4000B Scanner or equivalent).
5. 1 L Beakers (VWR)

2.8. Post-hybridization Analysis

1. Computer, preferably with 2 GB of RAM and software for analysis such as S-Plus, Matlab, or Spotfire.

3. Methods

3.1. Extraction of DNA from Small Number of Cells

When preparing DNA from small numbers of cells, the use of columns in kits can cause the loss of material. We have adapted a method that does not use columns.

1. Centrifuge 50,000–200,000 cells in 1.5 mL siliconized tubes in a microcentrifuge at 12,000g at 4°C. Pipette off the supernatant, and resuspend the pellet in 100 μL TE buffer (pH 8.0) containing 20 μg/mL tRNA by vortexing for 1 min. Centrifuge briefly to collect the liquid at the bottom of the tube.

2. Add 0.1 volume 10% SDS and 0.1 volume proteinase K, mix by gentle tapping and incubate at 55°C for 1 h. Tap periodically to accelerate digestion.

3. Centrifuge briefly to collect the liquid at the bottom of the tube. Add equal volume of phenol saturated with TE, mix by vortexing for 30 s, and centrifuge 5 min to separate the phases. Transfer the aqueous phase into a new siliconized tube and discard the organic phase. Repeat this with phenol:chloroform.

4. Precipitate DNA by adding 1/10 volume 3 M sodium acetate pH 5.2 and 2.5 volumes ethanol. Mix by inverting the tubes several times. Incubate at –20°C (or wet ice) for 15 min. Centrifuge in a microfuge at top speed for 15 min, and remove supernatant. Wash with 70% ethanol, centrifuge for 2 min, remove supernatant, and dry pellets under vacuum (*see* **Note 11**).

5. Dissolve each pellet in 50 μL TE pH 8.0.

6. Determine DNA concentration with Nanodrop.

3.2. Restriction of DNA

1. Digest DNA (1 μg if possible) in a total volume of 100 or 200 μL using at least 10 U (1 μL at 10,000 U/mL) of *BglII* in a 1.5 mL Eppendorf tube. Place tube(s) in a 37°C water bath and digest 6 h to overnight (*see* **Note 12**).

2. After digestion, centrifuge briefly to collect DNA at bottom of tube and add 1 μg tRNA (1 μL 10 mg/mL stock) and extract one time with equal volume of phenol:chloroform (*see* **Note 13**). Add the phenol:chloroform to the digestion mixture, vortex for 5 s, centrifuge at 16,100 rcf for 5 min. Remove and retain the aqueous upper layer, transferring to a clean Eppendorf tube. To that add equal volume chloroform:isoamyl alcohol, vortex, and centrifuge. Remove and retain aqueous layer, transfer to clean tube.

3. Add 1/10 volume 3 M sodium acetate pH 5.2 and 2.5 volumes ethanol. Mix by inverting the tubes several times. Incubate at −20°C (or wet ice) for 15 min. Recover DNA by centrifugation 15–20 min in a microfuge (unless otherwise mentioned assume top speed), remove supernatant and wash pellet with 70% ethanol, spin 2 min in microfuge, remove 70% ethanol, and dry in speed vac (*see* **Note 14**).

3.3. Ligation of Adaptors

1. To the pellet of DNA, add 7.5 µL each of the 12-mer (IBgl12) and 24-mer (IBgl24) oligonucleotides, 11 µL diH$_2$O and 3 µL 10X ligase buffer, and vortex.

2. To anneal the oligonucleotides, place the tube(s) into the thermal cycler and decrease the temperature as follows. Begin at 55°C and drop 1°C per minute to a 10°C hold (*see* **Note 15**).

3. Place tube(s) on ice for several minutes, add 1 µL T4 DNA ligase (10,000 U/µL) and mix by pipetting. Incubate overnight in thermal cycler set to a 15°C hold (*see* **Note 16**).

3.4. PCR Amplification and Clean-Up

1. To the ligation add 1 µL tRNA and 270 µL TE. Add 3 µL of ligation to each of eight PCR tubes. Prepare a cocktail of the PCR mix including 2 µL IBgl24 mer (desalted, not HPLC) with enough for all tubes plus 2 extra samples (to allow for pipetting error) (*see* **Note 17**).

2. Amplify in thermal cycler using the following program (*see* **Note 18**) and **Fig. 4.1**:

   5 min at 72°C

   20 cycles of:

      1 min at 95°C

      3 min at 72°C

   10 min at 72°C

   4°C hold

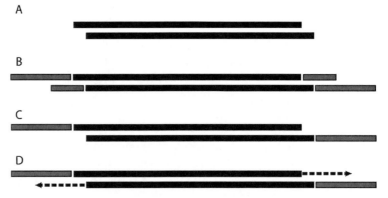

Fig. 4.1. Schematic illustration of the structure of the DNA primer hybrids. **(A)** Restricted DNA, **(B)** at 15°C during ligation (adapters shown in *dark gray*), **(C)** after ligation at 72°C, **(D)** extension during the first 5 min.

3. After PCR amplification, run 10 μL of PCR product on a 2% agarose gel, with size standards to make sure product was obtained. See **Fig. 4.2** for an example of what the representation should look like.

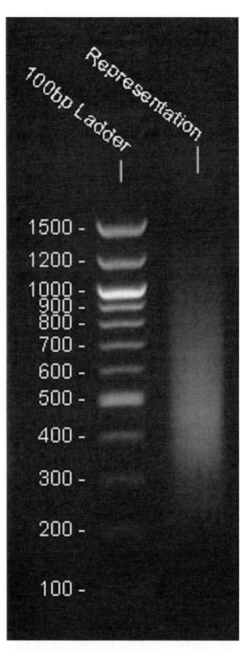

Fig. 4.2. Verification of generated representation by agarose gel electropheresis. The *right-hand lane* shows the expected characteristic smear of a representation. The marker in the *left lane* is a 100 bp ladder.

4. Combine contents of eight tubes (1 tube strip or one column of 96-well plate) into one 2 mL Eppendorf tube. Extract one time with equal volume of phenol:chloroform and one time with equal volume of chloroform.

5. Add 1/10 volume of 3 M sodium acetate pH 5.2 and 1 volume isopropanol, and incubate in a wet ice bath for 15 min to 1 h. Recover DNA by centrifugation (16,100 rcf (for 30 min) and wash pellet with 70% ethanol. Dry in speed vac and resuspend in TE at 1/10 the starting PCR volume.

6. The concentration can be determined by taking the OD; however, we utilize a Nanodrop ND1000 to decrease the amount of material lost.

3.5. Labeling Protocol for Microarray Hybridizations

1. Place 5 μg of DNA template for each sample (test and reference) in a 0.2 mL PCR tube. In general, two hybridizations are performed for each sample (commonly referred to as dye swaps) by changing the fluorescent dye labeling order. The ratio of the first hybridization and the inverse of the dye swap ratio are then averaged to remove color bias. This would make four labelings for each sample pair.

2. Add 5 μL primer solution and pipette up and down several times. Bring volume up to 40 μL with diH$_2$O. Mix gently and quickly spin.

3. Place tubes in thermal cycler at 100°C for 5 min (with heated lid). Quickly spin and immediately place on ice for 5 min.

4. Add the following:

   6 μL labeling buffer

   4 μL label (Cy3-dCTP or Cy5-dCTP)

   2 μL Klenow fragment

   6 μL dNTPs

   2 μL diH$_2$O

   Total volume ∼ 60 μL

   Pipette entire volume up and down and quickly spin. Incubate at 37°C for 2 h in thermal cycler (*see* **Note 19**).

5. Combine the labeled samples (Cy3 and Cy5) into one Eppendorf tube.

6. Add: 50 μL 1 μg/μL Human Cot-I DNA

   10 μL 10 mg/mL stock Yeast tRNA

   200 μL Low TE

7. Load all into a Microcon YM-30 filter and centrifuge at 12,600 rcf for 10 min. Check color of flow-through – should be deep purple (*see* **Note 20**).

8. Discard flow-through and wash with 450 μL Low TE. Centrifuge at 12,600 rcf.

9. Repeat this wash step two or three times, until flow-through is clear (*see* **Note 21**).

10. Invert the Microcon filter into a new Microcon tube. Centrifuge at 12,600 rcf for 2 min.

11. Transfer the sample into a 0.2 mL PCR tube. Check volume and adjust volume to 10 μL with low TE (*see* **Note 22**).

3.6. Pre-hybridization and Hybridization

1. Use Nimblegen array locator to demarcate array on the slide. Mark array with a diamond pencil on underside, being careful not to touch the array.

2. Prepare a beaker with deionized H_2O, wash slides with agitation for 30 s.

3. Immediately immerse slides in ice-cold 95% benzene-free ethanol (stored at $-20°C$) for 1 min.

4. Dry slide (*see* **Note 23**).

5. Store slides at room temperature in slide box prior to hybridization (*see* **Note 24**).

6. Prepare 35 mL hybridization buffer in order as follows (*see* **Note 25**):

Reagent	Volume	Final Concentration
Low TE	340 μL	–
100% formamide	1,750 μL	50%
20X SSC	875 μL	5X
10% SDS	35 μL	0.1%
diH₂O up to 35 mL		

Preheat hybridization buffer in a 37°C H_2O bath for 5–10 min.

7. Mix 30 μL hybridization buffer with 5 μL labeled sample and add 0.2 μL alignment control oligonucleotides.

8. Denature sample in thermal cycler programmed for 5 min at 95°C then linked to 37°C for at least 5 min but no longer than 30 min.

9. Once sample(s) have been at 37°C for 5 min, quickly get slides and place in hybridization tray. Place coverslips over array (with teflon side facing array) and Whatman filters (prewet with 800 μL 5XSSC) in hybridization tray.

10. Remove samples from thermal cycler, quick spin samples, and immediately pipette up and down gently to mix, then pipette under coverslip. Cover hybridization tray and place back into hybridization oven already at 42°C for 14–16 h (*see* **Note 26**).

3.7. Washing and Scanning

1. Remove slide(s) from hybridization oven, place in slide holder, and wash as follows (*see* **Note 27**):

   Quick dip in 0.2% SDS with 0.2X SSC to remove coverslip.

   1 min in 0.2% SDS with 0.2X SSC twirling slide rack

   30 s in 0.2X SSC twirling slide rack

   30 s in 0.05X SSC twirling slide rack

2. Dry slides (*see* **Note 23**). Scan immediately (*see* **Note 28**) and save the data as a tiff file (*see* **Note 29**).

3. The intensity information can be extracted from the tiff file by using NimbleGen software, presently called NimbleScan, and a text file can be exported. The software also adds probe identification information. Once exported, other genomic annotation can be merged to the data, such as genomic coordinates.

3.8. Post-hybridization Analysis

1. The first step of analysis is to normalize the intensity data. This can be done using several algorithms. We presently use a lowess curve-fitting algorithm (commonly used for data normalization) adapted from methods described by Dr. Terry Speed and colleagues *(8)*.

2. After normalization, the probe ratios can be calculated and averaged for the color reversals if performed.

3. Data can be analyzed with various software packages with capabilities for association of annotation and visualization of large datasets. To visualize the CGH profiles there are various possibilities. The simplest would be to use MS Excel, with the 2007 version allowing over 64,000 rows (a limitation of older versions). The most complicated would be to utilize the R language for all analysis and visualization which would require performing all commands by script. In between these two alternatives are software packages such as Matlab and S-Plus. Both these packages allow analysis and visualization. Data can be manipulated, annotated, merged, and visualized in various ways. Much of this can be done by writing scripts or using the graphical user interface.

4. To decrease or remove probe to probe variation, various algorithms can be used. A simple but effective way is to calculate a moving average, which moves in genomic order and averages adjacent probes. The number of probes being averaged can be adjusted. Improvements have been made to this algorithm *(4)*. There are also other more complex algorithms that identify regions with similar ratios and calculate a mean or median value. We utilize an algorithm developed by Adam Olshen and colleagues *(9)*, called Circular Binary Segmentations (CBS), that identifies regions of similar ratio

and then gives all probes in the segment the mean ratio. The code for this algorithm is freely available (http://bioconductor.org/packages/2.1/bioc/html/DNAcopy.html).

5. Analysis of larger datasets is frequently performed to identify common regions of amplification or deletion. If doing so, it is important to remove or at least demarcate the regions of the genome that frequently vary in normal individuals. The genome annotation for these regions can be found in several publications *(10, 11)*, and the identification from the dataset can either be done by hand or by writing a script to do so.

6. The common regions of alteration can be easily identified by calculating the frequency that a region is found above a threshold. Generally we find that a simple way of identifying alterations is to identify segment values (see Step 4 above) that are one standard deviation above or below for the particular experiment being analyzed.

4. Notes

1. Use HPLC grade purification when ordering the 24 and 12 mer for use in the ligation since it is a sensitive reaction. The 12 mer is not phosphorylated and will not ligate, but serves as a bridge to hold the 24 mer in place to allow ligation of the 24 mer.

2. The ligase buffer should have a DTT aroma; if not, it is old and should not be used in the ligation reaction.

3. Since the PCR is not as sensitive as the ligation reaction, desalted oligos can be used in place of the more expensive HPLC-purified oligos.

4. Make small aliquots of dNTPs to avoid multiple freeze–thaw cycles; thaw on ice.

5. Dyes, especially Cy5, are extremely light sensitive so work in low-light conditions.

6. Use Cot for appropriate species (e.g., human or mouse). This serves mostly to remove possible bridging between repetitive sequences that can increase background hybridization.

7. tRNA is used to decrease the slide background signal.

8. Dyes are sensitive to high Tris concentrations.

9. Detailed description of probe selection and array design is available from two references *(5, 12)*. Briefly, all probes on the array are designed to be complementary to representational fragments, i.e., small restriction fragments. Probe design is

first started by making an in silico representation based on the available genome sequence. This is performed simply by locating and recording all instances of the restriction site of interest and size-selecting to 100–1,200 bp. A Burrows–Wheeler transformation *(13)* was used to identify the number of times that any region of a fragment is found within the genome. Probes were selected with the least repeat content and then tested empirically. We typically begin by designing and arraying tenfold more probes than can fit on one array. We select those probes with good performance. Finally the placements of probes on the array surface are randomized. In this way, no hybridization artifact on the array surface will appear as a consistent genomic alteration. The probe sequences can be devised from the probe coordinates from the following website (http://lucitolab.cshl.edu/rl_data.html) using the columns CHROM, FRAG.POS, and PROBEID. All coordinates are for position one of the 50 mer. Using these coordinates and extending 50 nt will yield all probe sequences.

10. Labeled alignment oligonucleotides are light sensitive and subject to degradation after multiple freeze–thaw cycles. Therefore, prepare 5 µL aliquots and store in light-protective tubes at −20°C.

11. If the DNA is left to dry for too long it will be hard to resuspend.

12. If a different representation is being used, be sure to use the corresponding restriction endonuclease chosen for the array design.

13. The Qiagen DNA purification kit can be used to purify DNA instead of using the phenol cleanup as described here. A disadvantage is that more DNA is lost as compared to the phenol-based method.

14. Inspection of the digest by agarose gel electrophoresis can be performed if desired, adjusting the reaction volume in the subsequent step. If small amounts of DNA are digested, the digestion may not be visible unless compared to undigested DNA. We will use for simplicity the analysis of cancer as an example. Be sure to digest a reference sample alongside the test sample. The reference preferably is from the same individual but can be unrelated.

15. Instead of using a thermal cycler, tubes can be placed in a 55°C heat block with removable block. Remove the block and place in a cold room or refrigerator until the block temperature reaches 10–12°C.

16. Alternatively, place tubes in a 15°C water bath.

17. Usually eight reactions per sample yields more than enough representation for experiments.

18. It is imperative to start with 72°C and not 95°C due to the structure of the hybrids. If 95°C is used first, no representation will be amplified. 72°C is required to allow filling in of the fragment ends to create the primer binding site for subsequent cycles of PCR (*see* **Fig. 4.1**). If a large number of samples are being analyzed, the amplification can be performed in 96-well format. For digestions under 200 ng, amplify for 25 cycles.

19. Dyes are light sensitive, especially Cy5, so work in low-light conditions. Do not vortex tube containing Klenow fragment, keep on ice. Can incubate up to 6 h; 2–3 h works best. At this time, pre-scan slides to check for any defects in the slides (scratches, unevenness).

20. Use best judgment with color. Sometimes flow-through is not always a dark purple, may be lighter. Do not spin filter unit at max speed; it will destroy the column! If sample completely washes through filter try using flow-through in another filter. It could be a faulty filter.

21. If sample completely washes through the column, it is likely that the sample did not label properly. Try repeating the labeling protocol (**Section 3.5**). If the same results are obtained, remake representations, and repeat (starting at **Section 3.4**).

22. If sample is more then 10 μL, put back into Microcon filter with Low TE and centrifuge, then invert into new tube, and centrifuge for 2 min.

23. Air blowing of slides is accomplished by using a steady flow of air from bench-top outlets with filter. Do not use canned gas; if not careful, liquid comes out which will ruin hybridization to the slide. An alternative, if performing many hybridizations, is to use compressed nitrogen which can be obtained from companies supplying CO_2 for tissue culture.

24. Avoid dust as much as possible. All wash volumes are ~500–600 mL. Slides must be dried immediately after ethanol wash. CyDyes are extremely light sensitive. Minimize exposure to light by carrying out all experimental procedures in low-light conditions.

25. To reduce variability, prepare a large master mix of at least 100 times volume. A large volume of hybridization buffer is prepared to allow for manageable volumes of buffer constituents and to decrease pipetting error.

26. Can hybridize longer but not recommended as this increases background.

27. Wash volume is 500–600 mL in 1 L beaker.

28. Cy dyes, especially Cy 5, are very sensitive to ozone levels over 5 ppb and this is the worst in the summer. The average levels can be checked on national weather websites. Scanning early in the morning when ozone is at its lowest seems to alleviate this problem.

29. Presently we use an Axon GenePix 4000B Scanner but other scanners are adequate such as Agilent or Perkin Elmer. It is important that the pixel size for scanning be 5 μm.

References

1. Aguirre, A. J., Brennan, C., Bailey, G., Sinha, R., Feng, B., Leo, C., et al. (2004) High-resolution characterization of the pancreatic adenocarcinoma genome, *Proc. Natl. Acad. Sci U. S. A.* **101**, 9067–9072.

2. Barrett, M. T., Scheffer, A., Ben-Dor, A., Sampas, N., Lipson, D., Kincaid, R. et al. (2004) Comparative genomic hybridization using oligonucleotide microarrays and total genomic DNA, *Proc. Natl. Acad. Sci. U. S. A.* **101**, 17765–17770.

3. Pinkel, D., Segraves, R., Sudar, D., Clark, S., Poole, I., Kowbel, D., et al. (1998) High resolution analysis of DNA copy number variation using comparative genomic hybridization to microarrays, *Nat. Genet.* **20**, 207–211.

4. Pollack, J. R., Perou, C. M., Alizadeh, A. A., Eisen, M. B., Pergamenschikov, A., Williams, et al. (1999) Genome-wide analysis of DNA copy-number changes using cDNA microarrays, *Nat. Genet.* **23**, 41–46.

5. Lucito, R., Healy, J., Alexander, J., Reiner, A., Esposito, D., Chi, M., et al. (2003) Representational oligonucleotide microarray analysis: a high-resolution method to detect genome copy number variation, *Genome Res.* **13**, 2291–2305.

6. Lisitsyn, N., Lisitsyn, N., and Wigler, M. (1993) Cloning the differences between two complex genomes, *Science* **259**, 946–951.

7. Lucito, R., Nakimura, M., West, J. A., Han, Y., Chin, K., Jensen, K., et al. (1998) Genetic analysis using genomic representations, *Proc. Natl. Acad. Sci. U. S. A.* **95**, 4487–4492.

8. Yang, Y. H., Dudoit, S., Luu, P., Lin, D. M., Peng, V., Ngai, J., et al. (2002) Normalization for cDNA microarray data: a robust composite method addressing single and multiple slide systematic variation, *Nucleic Acids Res.* **30**, e15.

9. Olshen, A. B., Venkatraman, E. S., Lucito, R., and Wigler, M. (2004) Circular binary segmentation for the analysis of array-based DNA copy number data, *Biostatistics* **5**, 557–572.

10. Iafrate, A. J., Feuk, L., Rivera, M. N., Listewnik, M. L., Donahoe, P. K., Qi, Y., et al. (2004) Detection of large-scale variation in the human genome, *Nat. Genet.* **36**, 949–951.

11. Redon, R., Ishikawa, S., Fitch, K. R., Feuk, L., Perry, G. H., Andrews, T. D., et al. (2006) Global variation in copy number in the human genome, *Nature* **444**, 444–454.

12. Healy, J., Thomas, E. E., Schwartz, J. T., and Wigler, M. (2003) Annotating large genomes with exact word matches, *Genome Res.* **13**, 2306–2315.

13. Burrows, M., and Wheeler, D.J. (1994) *in* "Technical Report 124," Digital Equipment Corporation, Palo Alto, CA.

Chapter 5

Application of Oligonucleotides Arrays for Coincident Comparative Genomic Hybridization, Ploidy Status and Loss of Heterozygosity Studies in Human Cancers

John K. Cowell and Ken C. Lo

Abstract

Many oligonucleotide arrays comprise of spotted short oligonucleotides from throughout the genome under study. Hybridization of tumor DNA samples to these arrays will provide copy number estimates at each reference point with varying degrees of accuracy. In addition to copy number changes, however, tumors often undergo loss of heterozygosity for specific regions of the genome without copy number changes and these genetic changes can only be identified using arrays that identify polymorphic alleles at each reference point. In addition, because the hybridization intensity can be measured at each of the allelic variants, allelic ratios can be established which give indications of ploidy status in the tumor which is not generally possible using most other oligonucleotide array designs. The only arrays currently available that simultaneously report copy number, ploidy, and loss of heterozygosity are the Affymetrix SNP mapping arrays.

In this review, the features of the SNP mapping arrays are described and computational tools explored which allow the maximum genetic information to be extracted from the experiment. Although the methodologies to generate the SNP data are now well established, approaches to interpret the data are only just being developed. From our experience using these arrays, we provide insights into how to evaluate the SNP data to report copy number changes, loss of heterozygosity, and ploidy in the same tumor samples using a single array.

Key words: SNP mapping arrays, comparative genome hybridization, loss of heterozygosity, allelic ratios, CGH visualization tools, oligonucleotide arrays.

1. Introduction

Significant genetic events promoting tumorigenesis involve structural and numerical chromosome changes, such as losses, gains, and amplifications, that result from copy number abnormalities

Jonathan R. Pollack (ed.), *Microarray Analysis of the Physical Genome: Methods and Protocols, vol. 556*
© Humana Press, a part of Springer Science+Business Media, LLC 2009
DOI 10.1007/978-1-60327-192-9_5 Springerprotocols.com

(CNAs). Characterization of these events can identify consistent changes in the cancer genome that are potentially related to transformation and progression. Independently, these CNAs could also serve to predict profiles related to response to therapy (1) and facilitate molecular classification and risk assessment (2). Consistent deletions suggest the location of tumor suppressor genes, and high-level amplifications often identify oncogenes. The exposure of recessive mutations in tumor suppressor genes, however, can also occur in the absence of CNAs and result from mitotic recombination and chromosome non-disjunction, for example (3), and manifest in loss of heterozygosity (LOH) on an otherwise copy number neutral (CNN) background. To identify these CNAs at high resolution, a variety of microarray platforms have been developed for comparative genome hybridization (aCGH). These arrays carry probes derived from either cDNAs, large insert clones such as bacterial artificial chromosomes (BACs), or oligonucleotides with varying designs (4). All of these arrays will identify CNAs with varying resolution and accuracy, but the Affymetrix Gene mapping arrays will simultaneously report LOH and CNAs, as well as estimate ploidy levels in the same experiment. This multiple capability is a feature of the array design, which contains targets that interrogate single-nucleotide polymorphisms (SNPs) in the human genome and so can establish allelotypes at each reference point. For this reason the details in this article focus specifically on the SNP mapping array (SMA) platform from Affymetrix.

2. Hybridization to SMAs

Probably the most important aspect of the SMA analysis is the quality of the DNA used to generate the labeled target for the hybridization reaction. Specifically, Nannya et al. (5) demonstrated that larger product sizes may not be amplified if the template DNA is partially degraded. This may be particularly true for DNA derived from formalin-fixed paraffin-embedded samples, for example. In our experience also, some 'karyotypes' generated by the SMA procedure are noisier than others, and agarose gel analysis of the template DNA used to generate the probe shows that those with any evidence of degradation tend to be more noisy (see later). The level of contamination of the tumor sample by normal cells (which is clearly not an issue with cell lines) can contribute to the noise of the arrays, as well as the heterogeneity within the tumor cells themselves, where different karyotypes in different subclones will alter the profile to varying degrees depending on their overall representation within the tumor. This follows since the CGH analysis reports the net consequences of the sum total changes in the tumor which can potentially cancel each other out. The consequences of these complications are manifested in the analysis of the data, however, and do not affect the processing of the arrays. For most users, the SMAs will be processed by core service facilities which follow the recommended Affymetrix protocol depending on which specific SMA is being used (see www.affymetrix.com). The principle of this procedure is outlined below:

1. ~250 ng of high-molecular-weight DNA is digested with the appropriate restriction enzymes that match the specific array.

2. An adaptor specific to each reaction is then ligated to the ends of the DNA fragments.

3. A PCR is then performed using primers directed against the adaptor sequences.

4. The PCR products are then purified and fragmented using DNase I.

5. The fragmented products are end-labeled with biotin and hybridized to the chip.

The different enzymes used to generate the probes are specific to the arrays, which have been designed to ensure that the genomic targets generated by PCR correspond to the SNPs on the arrays. At the time of writing, 100 K and 500 K arrays are in common use, but the Affymetrix platform at this time is transitioning to the SNP 6.0 array which contains 1.8 million reference points on a single chip. While the overall protocols remain the same, the specific details for each series of chips is different and it is advisable to discuss the type of array available and the concentration of DNA needed with the Core facility or provider to ensure the correct matching of target and array. In terms of the data analysis, and the identification of events within the cancer genome, we have found that the smallest, 100 K array will report most CNA and LOH events in a given sample, although the 500 K array potentially gives greater resolution and improves LOH statistics. If cost is important, then for the most part, the data generated from a single 50 K or 250 K array will also define most of the CNA and LOH events in the karyotypes (see later for discussion of considerations for computing power for analysis of this data). In a recent survey of the same tumor DNAs using several different array platforms, we identified relatively few additional CNAs using the 500 K arrays which, for us, had no particular cost–benefit in the study (6–7).

3. Data Analysis and Interpretation

3.1. Data Processing

The Affymetrix-supported CNAT4.0 suite or Affymetrix Power Tools (APT) undertakes all of the processing, normalization, and reporting of the CNAs and LOH events. After processing, there are two defining characteristics that can be used to evaluate the quality of the dataset: the 'call rate' percentage and the sample $\log_2$ ratio inter-quartile range (IQR). The call rate defines how many of the SNPs were assigned a genotype based on the relative differences between the intensity values from the major/minor alleles. Although a call rate of >96% is expected for most experiments, lower call rates do not necessarily mean that the experiment has failed but rather that there was likely something wrong with the DNA or its processing. In our experience, call rates of 90–95%, and even some in the 85–90% range, provide very acceptable karyotypes. Significantly, lower call rates should be viewed with caution and technical errors should be suspected. Repeating the experiments in these cases will potentially solve the problem, unless it was caused by poor-quality DNA used to prepare the labeled target.

Overall, but depending on the objective of the experiment, plotting the aCGH karyotypes is an important first step, since low call rates may still be usable where only large events, such as chromosome arm losses associated with clinical phenotypes, need to be identified. As a discovery tool, however, it is likely that loss of signal for 15–20% of the points on the array may lessen the chances of identifying small events in the karyotypes. However, for samples where the DNA was limiting, such as from fine-needle biopsies, for example, repeating the experiment may not be an option. Once an acceptable call rate has been established, the processing of the data can proceed.

3.2. Data Analysis Considerations

As a starting point, the compendium of tools that is included in the Affymetrix Copy Number Analysis Tool 4.0 (CNAT 4.0) suite provides numerous visualization and analysis options for the end user and is driven by a graphical user interface (GUI). Command line versions of the same programs are also available as the APT suite. The following discussion applies to the 'unpaired' workflow, which uses the allelic frequencies of a group of selected HapMap individuals to serve as a 'control' group that is provided at the Affymetrix website (although reference sets can also be custom generated to reflect the population under study more closely).

3.3. Copy Number Analysis Options

3.3.1. Smoothing

Due to the signal-to-noise ratio of the Affymetrix GeneChip platform, raw summarized intensities are generally not helpful in identifying copy number gains and losses, and thus, the default settings use a Gaussian smoothing algorithm set with a bandwidth of 100 kb. The specific advantages/disadvantages of this smoothing is explored in more detail in the Affymetrix Copy Number Analysis Tool (CNAT) 4.0 Workflow guide. Simplistically, the smoothing process helps in identifying events that are larger than 100 kb but dampens the true signal for events smaller than 100 kb. As a starting point, we perform a general inspection of the overall noisiness of each of the arrays by plotting the Gaussian smoothed $\log_2$ ratios on aCGHViewer (8), which separates each chromosome in an individual panel to explore it in greater detail. We have found other tools, which tile each chromosome end-to-end, tend to obscure some important information, such as the greater noise level on chromosome 19 on the 100 K SMA, for example.

3.3.2. Restriction by Fragment Size

One option in the 'Advanced Analysis Options' allows the analysis of the SNPs on the array to be restricted to the sizes of PCR products generated during the amplification step. An advantage of this analysis was first explored in Nannya et al. (5), where they observed poorer quality DNA generally does not give the same range of PCR products during the amplification step. By restricting the analysis to SNPs that are amplified in fragments within specific size ranges, SNPs from larger PCR products are removed if there is an a priori suggestion that the amplification of larger products is limiting.

3.3.3. Normalization

Normalization allows a more comparable analysis of data from different samples. Quantile normalization is the default setting, which performs a sketch quantile normalization on the perfect match (PM) probes across the samples by enforcing the empirical distributions of the PM probe intensities in each of the samples to be the same. While there is still debate about the fundamental assumption that the probe intensity distribution for different samples should be the same, it has nonetheless become the default choice for normalization for many of the Affymetrix GeneChip products, since it provides a less biased way of making comparisons between samples.

3.4. Genome Segmentation Approaches

The Hidden Markov Model for identifying transition points in DNA copy numbers was first explored by Fridlyand and colleagues (9). While this approach has been popular, the original segmentation algorithm used in this approach has since been extended by others and alternative algorithms, such as circular binary segmentation (CBS), GLAD, and others (reviewed in (10)) have been shown to outperform the original HMM in terms of identifying segment boundaries when the signal-to-noise ratio is low. With the default parameters, we have found that the five-state HMM that is included with CNAT 4.0, results in oversegmentation of the data in what are, for example, clear examples of regional loss along a chromosome arm. While we have not fully explored the tunable parameters of the HMM incorporated into CNAT 4.0, we have adopted the CBS algorithm for segment calling, and have found it to be more consistent in terms of identifying large segments in our application. A note of caution with the CBS segmentation algorithm, however, is that Gaussian smoothing removes the variance in the raw log_2 ratios and, since the CBS algorithm uses the variance measure for statistical testing, many small segments will be produced. We have thus adopted the following method of segmentation:

1. Produce raw log_2 ratios (either from CNAT 4.0 or APT) by setting Gaussian smoothing to 0.

2. Format the data to be consistent for the R library 'DNAcopy'.

3. Segment the genome using default settings, with no trimming of the segments.

4. Call each segment by using $0.3x$ MMAD of each genome: outliers are detected using $2x$ segment IQR for flagging, $2x$ global IQR for calling.

5. Combine the segment calls with the Gaussian smoothed data for display.

3.5. Copy Number Abnormalities

Probably the most extensive, and well-characterized, category of genetic abnormalities in tumors involves structural and numerical chromosome changes which are related to the development and progression of cancer (11). aCGH will detect a CNA where there has been a loss or gain of chromosomal material. Reciprocal chromosome translocations, however, will not be detected.

One of the features of oligonucleotide arrays is that, based on the relative intensity of hybridization of the probe at each locus, estimates of copy number can be determined relative to reference standards. This analysis makes it possible to determine copy number changes along the length of each chromosome. The ability to clearly define the CNAs depends to some extent on the noise in the dataset (see above). Where noise is low, the profiles are relatively easy to interpret, but where the noise is high, CNAs may be less obvious and so segmentation algorithms such as CBS become particular important. After processing, individual segments are defined as losses, gains, or no change and these are then plotted in the preferred viewer to generate a whole-genome view of the karyotype or on a chromosome-by-chromosome basis.

As shown in **Fig. 5.1**, whole-chromosome losses are relatively easy to identify and generally show a log_2 ratio close to the theoretical single copy loss value of -0.5. Values between -0.4 and -0.5 are more usual, however, depending on the noise in the sample. Similarly, single-copy gains are revealed with a log_2 ratio which similarly do not achieve the theoretical maximum of $+0.58$ (**Fig. 5.1**). Where gains, losses, or 'no changes' (log_2 ratio $= 0$) are identified along the same chromosome, the transitions are also

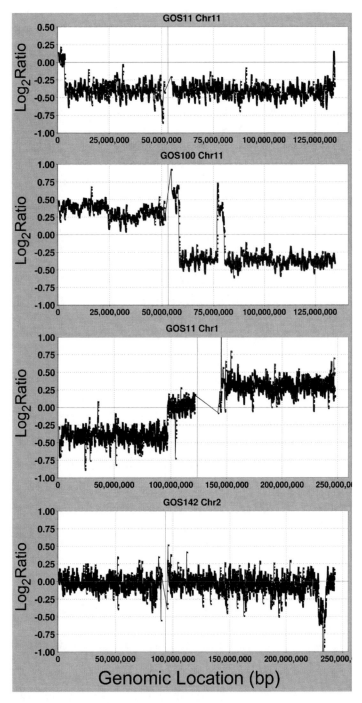

Fig. 5.1. Examples of aCGH profiles from SMAs. Loss of the majority of chromosome 11 (*top*) demonstrates a log$_2$ ratio of ~-0.4 except for the distal tip of 11p which shows normal diploidy and a log$_2$ ratio of 0. Complex CN events are shown for chromosome 11 (*upper middle*) with gain of the short arm and loss of the long arm. In this case, amplification events are also seen on the retained copy of the long arm. Examples of losses, gains, and no change on chromosome 1 (*lower middle*) are clearly seen in context with each other. Homozygous deletion in distal 2p (*below*) with a log$_2$ ratio of >-1 is seen on a background of hemizygous deletion for the flanking region.

relatively clear (**Fig. 5.1**). Although single copy losses are seen with an approximately -0.5 $\log_2$ ratio, homozygous deletions should theoretically show no signal. In practice, however, we have found that because of the presence of normal cells in the population, these deletions show ratios between -1.0 and -2.0. However, in context with hemizygous deletions, as shown in **Fig. 5.1**, these homozygous deletions are relatively easy to identify, although these observations would usually need some independent verification, such as FISH analysis as we have reported previously *(12–13)*. Single point events should be viewed with suspicion, but where multiple adjacent reference points are involved, there can be more confidence that the observation is real.

The position of the transition from no change to loss or gain can usually be identified directly from the viewer, if that function is available (as in *(8)*), or, alternatively, through analysis of the raw data files where, in CNAT 4.0, the base pair location along the length of the chromosome of the individual SNPs is defined, and the copy number reported. The definition of the exact location of the breakpoint, however, is only as accurate as the density of the SNPs on the array in that location, and often the transition point is not clearly defined by two adjacent SNPs, since the smoothing algorithm used in CNAT 4.0 at these points tends to smooth out transitions needed to make an accurate base pair identification (see earlier). Unless some other means of defining the exact breakpoint is available, therefore, we have adopted the convention of defining the breakpoint as starting and ending at the flanking SNPs that belong to segments of different categories. In this way the maximum extent of the CNA is defined, which can be important in attempts to include all of the potential genes that are associated with the CNA.

Although the terminology describing amplification has been used liberally, it is important to understand the underlying biology. Traditionally, amplification has been used to describe highly selected events that involve specific increases in intra-, or extra-chromosomal genetic material, rather than simple gains of whole chromosomes or chromosome regions *(14–15)*. Tumor cells frequently show gains of one or two copies of a particular chromosome segment, which results from non-disjunctional events at mitosis, in contrast to our definition of amplification. We have suggested previously, therefore, that if the increase in $\log_2$ ratio is no greater than twofold, then it should perhaps be described as a gain, with only high-level, focal increases in DNA content being reserved for the term 'amplification'. With this in mind, the SMAs tend to underestimate the magnitude of the amplification events (**Fig. 5.1**), possibly because of dye saturation, compared with, for example, BAC arrays *(6)*. Thus, on SMAs, large amplification events, such as the ones seen for the EGFR locus in gliomas, may only show a $\log_2$ ratio between $+2$ and $+4$ (**Fig. 5.1**).

3.6. Loss of Heterozygosity

LOH was proposed as a means of 'exposing' recessive mutations in tumor suppressor genes. Al Knudson's model *(16)* predicted that both copies of a tumor predisposition gene should be inactivated for tumorigenesis, which was subsequently demonstrated by Web Cavenee and colleagues *(3)*. There are several mechanisms whereby LOH can be detected using SMAs.

1. Loss of whole chromosomes or chromosome regions (CNA_{loss}) on a diploid background will invariably generate LOH.

2. Mitotic recombination events between sister chromatids, with the appropriate chromosome segregation, will generate LOH without CNA_{loss} – so-called copy number neutral events.

3. Chromosomal non-disjunctional events will also generate LOH without CNA_{loss}.

Since the SMAs can describe allelotypes, it is possible to distinguish between homozygosity and heterozygosity at each locus on the array. To define regions of LOH by identifying transitions between regions of heterozygosity (0) and homozygosity (1), the two-state HMM using default parameters (transition decay 10 Mb) generally performs well. To enhance visualization, we calculate an 'LOH score' by multiplying the homozygous allelic frequencies of the reference samples for a contiguous stretch of homozygote calls determined by the HMM model, which represents the odds that this stretch of homozygous alleles can occur by chance alone. Traditionally, the demonstration of LOH would require a comparison between the allelotypes in tumor and normal samples from the same individual, which would more accurately assess LOH. In many cases, however, normal material may not be available for these studies.

After processing, the data can be visualized as the $-Log_{10}(p\text{-val})$ for each segment for each sample (**Fig. 5.2**). The interpretation of these LOH scores are explored in more detail below. In the final analysis, determining whether LOH has occurred is related to the value of the LOH score. The accuracy of the determination, however, depends on the accuracy with which the individual alleles are identified as homozygous/heterozygous and the extent to which the allele frequencies in the reference set are representative of the group under study. Our data suggest that, at face value, the indication of homozygosity can sometimes be overstated in some circumstance *(6)* and so defining the threshold that accurately calls LOH is important (see below).

3.7. LOH Profiles

Regardless of the preferred visualization tool (reviewed in *(17)*), the identification of homozygosity is relatively straightforward. For example, the X chromosome from tumors from males should always be homozygous, even if normal cells are present, since they will also be XY *(6)*. If this is not the case, errors with processing, analysis, or input DNA quality should be suspected. In our experience, there are

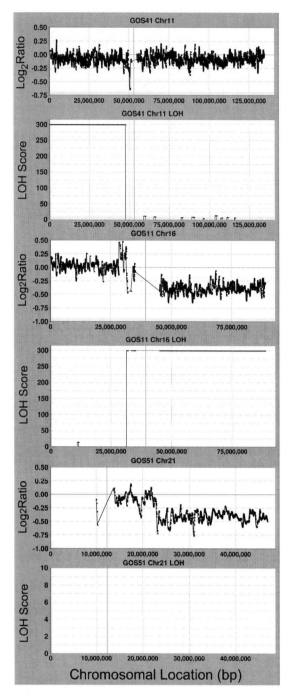

Fig. 5.2. Co-linear analysis of CNAs and LOH. (*Top*) Scores of ~300 on the short arm of this chromosome indicate LOH has occurred in the absence of CNA. CNA$_{loss}$ on the long arm of this diploid chromosome is associated with LOH (*middle*) as expected. CNA$_{loss}$ for the distal long arm with no associated LOH (*below*) indicates a tetraploid background.

relatively few calls of heterozygosity in XY samples, although in our experience some SNPs located within the pseudo-autosomal region, may frequently be called heterozygous in males *(6)*.

Where LOH is associated with CNA_{loss} there is usually no question about the validity of the observation. The challenge, however, is to determine whether LOH events seen on the LOH profile in the absence of CNA_{loss} are real or not. When viewing LOH profiles for the first time, there is a tendency to focus on the large spikes within the chromosomes that suggest minute regions of LOH (**Fig. 5.3**), even in the absence of CNAs for

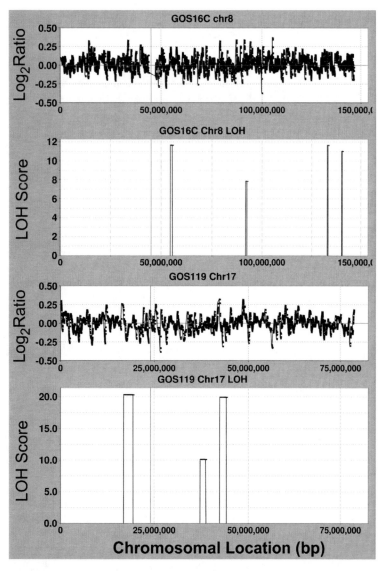

Fig. 5.3. Examples of apparent interstitial LOH without CNA_{loss}. In these two examples, CNN observations sometimes correspond to punctate regions of the chromosomes with LOH scores >10 and CN log_2 ratios clustering around 0 (no change).

the same region. Accepting these events at face value, however, will identify many apparent LOH events in the cancer genome, and there is a tendency to report these after setting an arbitrary threshold for the LOH score. A moment's reflection on how these events might occur, however, will generally demonstrate that they are probably artifacts, unless rigorously verified (see below). In an early release of CNAT (v3.0), an LOH threshold level of 10 was suggested, but we have shown by sequencing tumor/normal pairs within these regions of apparent LOH (6) that in many cases the patients were in fact constitutionally homozygous throughout the region (see above for details of the origin of the allele frequencies). Because of this consistent finding, we have raised the threshold in our analysis to >20, if the LOH is punctate as shown in **Fig. 5.3**, although there may still be examples where even this value may not represent LOH. It is at this point that a more intuitive interpretation may be necessary since, for multiple LOH events to have occurred along a chromosome arm in the absence of CNA_{loss}, multiple mitotic recombination events should have occurred, which is unlikely. We have failed to confirm these events as LOH in a tumor/normal comparison, no matter what the LOH score predicted (6). We speculate that the reference set used to calculate the allele frequencies may not always be representative of their frequency in the population under study. What was consistent, however, was that if whole chromosomes, or chromosome arm, showed high LOH scores (**Fig. 5.2**), then LOH was always confirmed after sequencing (6), and these events were due to either non-disjunction or consequences of chromosome translocations. In addition, where LOH was seen to be sub-regional, and extended to the respective telomere on the chromosome arm from a single transition point (**Fig. 5.2**), these events were verifiable by sequencing and could be interpreted as due to single mitotic events at the point of transition from heterozygous to homozygous.

From our experience with this LOH analysis, some general observations are:

1. LOH events with high scores, e.g. >50, can generally be accepted as real events since this is almost always the case where they are associated with CNA_{loss}.

2. Small interstitial regions of apparent homozygosity, in the absence of CNAs, would require double recombination events to occur and, if there are large numbers of these events in the karyotype, then they are probably not real, and in our experience show LOH scores of between 10 and 30. If these events are in regions of high relevance to the question being posed, however, then it is advisable to

validate the observation by some independent means such as reference to a normal sample from the same patient as we recently described *(6)*.

3. Although multiple, punctate LOH events are most likely not real, we have noted that in some cases where CNA_{loss} is clearly seen, the LOH score, although not uniform along the length of the chromosome arm, still shows many adjacent LOH segments with scores >30 (**Fig. 5.3**). Our interpretation of these profiles is that they represent LOH for the whole chromosome arm and that the rare, heterozygous calls along the length of the chromosome arm that break up the profile represent artifacts that are possibly due to contaminating normal cells in the sample.

4. No homozygosity in regions of CNA_{loss} usually indicates that the tumor is >diploid (see later).

5. Despite comments in (2) above, regions of interstitial LOH may represent selective amplification of a single allele at that locus, but this will be evident from the aCGH analysis (**Fig. 5.4** and see below).

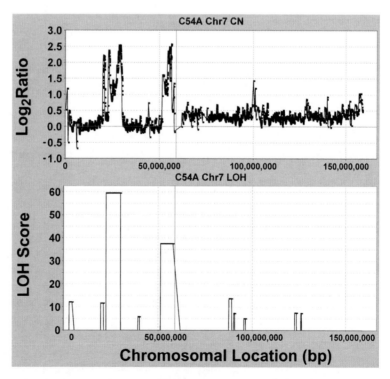

Fig. 5.4. Comparison of aCGH and LOH profiles for regions of amplification on the short arm of chromosome 7. Monoallelic amplification in 7p21 and 7p22 is demonstrated by the high LOH scores (>30) seen for these regions using the SMAs.

3.8. Defining the Position of a Breakpoint in an aCGH Karyotype

The position of the CNA-associated breakpoint, in addition to the identification of a CNA, may have critical importance to the study. For example, if a chromosome translocation is involved, then genes at the breakpoint may be affected (18), and so accurate determination is important. In addition, when defining deletions and amplicons, it may be important to accurately delineate the precise consequences of this event to correlate with gene expression changes (6, 19). With this in mind, it is important to consider the power of the individual array. Using lower-density arrays, the resolution is limited by the spacing of the SNPs on the array. Thus, even though the average representation may be quoted as 10 kb, because of unequal distribution along the length of the chromosome, due to design restrictions, these may, in fact, be more dense in some areas and less dense in others, or more dense on some chromosomes and less dense on others. In defining amplicons, gains, and deletions, therefore, it is probably necessary to define the maximum extent of a CNA if the borders of the transition from copy number neutral to copy number change are not discrete. In fact, in most cases this will not be the case because the Gaussian smoothing algorithm incorporated into CNAT4.0 will try to smooth transition points, in reality making the transition gradual.

Defining breakpoints associated with CNN LOH is more difficult since if the region adjacent to the breakpoint is constitutively homozygous, then the length of the LOH region will be extended proximally from the mitotic recombination event to the first constitutionally heterozygous locus in this individual. Thus, breakpoints in this analysis may be defined some distance from the actual event (6).

3.9. Characterization of Ploidy

One of the shortcomings of aCGH analysis, on all platforms, is that it will not generally predict ploidy, only changes in relative intensities of signal. In a comparison with normal cells, which are diploid, since a fixed amount of DNA is applied to the array, if, in its simplest form, the test DNA sample comes from a tetraploid tumor, then the same amount of DNA is applied, although it will be derived from only half the number of cells. Under these circumstances there will be no difference between two or four copies of a normal chromosome. If there are losses or gains on either ploidy background, however, CNAs will be detected because a shift in the $\log_2$ ratio is seen (6). Although for many applications this may not matter, if the intent is to define losses, gains, and amplifications, when the analysis incorporates LOH, this shortcoming can be profoundly important, since loss of a chromosome region on a tetraploid background creates a triplody that will still retain heterozygosity and not expose recessive mutations (see below). For this evaluation, there are several ways that the SNP arrays can be used to identify these events. Although still not perfect, these capabilities go a long way towards determining the ploidy of tumor cells, where a visual comparison of LOH and CNA events reveals two things:

1. If the tumor CGH profile, as a whole, shows numerous regions with CNA_{loss}, which also show LOH, then the tumor is almost certainly mostly diploid.

2. If none of the CNA_{loss} events show LOH, then it is likely that the tumor cells were tetraploid or higher.

It is often the case, however, that within a given karyotype, there will be examples where CNA_{loss} is associated with LOH, and others where they are not. This profile probably means that the events that show LOH occurred before the endoreduplication event that created the tetraploidy *(6)*, and the ones where CNA/ LOH did not correlate, occurred after this event. All of these correlations also hold true for subregional parts of chromosomes.

3.10. Determination of Ploidy Using Allelic Ratios

Although, as we have described above, the qualitative correlations between aCGH and LOH can go a long way towards determining diploid or higher ploidy status in tumors, the specific number of chromosomes present cannot always be determined based on the log_2 ratios alone. The primary restrictions for accurate ploidy analysis in this case are probably the experimental protocol, which calls for a standard amount of DNA to be assayed, as well as a result of the normalization procedures. Such procedures will approximately equalize the total signal regardless of whether the original cells were diploid or, for example, tetraploid. In addition, analytical procedures tend to normalize the overall signal approximate to those found in the reference population, which is almost exclusively diploid. As a result, computational methods relying on the total DNA signal will tend to grossly underestimate the actual copy number of tetraploid samples, since the presumptive baseline signal is set at 2 rather than 4.

Recently, we have developed an approach which analyzes the relative signal intensities for each allele of the individual SNPs along the length of the chromosome, and an in-depth consideration of this approach is provided in Gardina et al. *(7)*. The SMAs contain short oligomers to probe the alternative alleles of the SNPs throughout the genome. Furthermore, the signals from the alternative alleles can be segregated and translated into specific genotypes. The allelic ratio (AR) generates patterns that are characteristic of particular classes of CNAs, and these patterns can be distinguished from that of heterozygous diploidy. Essentially, the AR measures the contribution of the 'B' allele to the overall signal intensity at individual SNPs from the two possible alleles (A or B). Since the labeling of the allele is arbitrary, the ratios form a symmetrical pattern about a value of 0.5 within a region spanning equivalent copy number (CN) status. For a heterozygous, diploid chromosome, the ARs should approximate to either 1.0 (BB), 0.5 (AB), or 0 (AA), generating the pattern seen in **Fig. 5.5A**. The

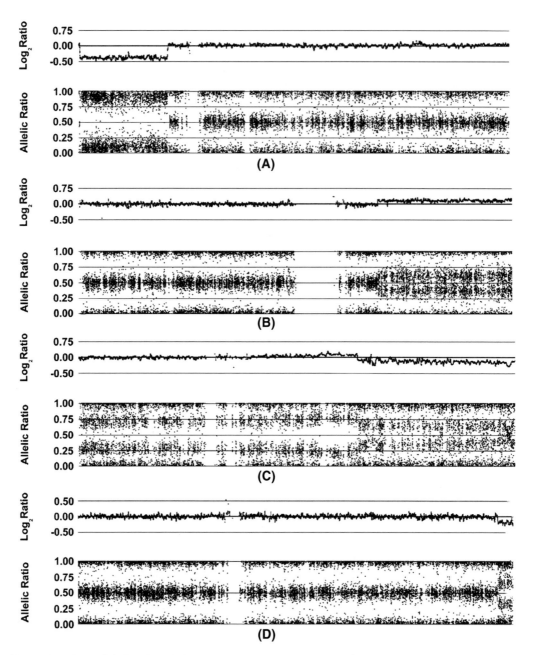

Fig. 5.5. Description of allelic ratio analysis in comparison with copy number. In (**A**) the long arm and proximal short arm of this chromosome is diploid and shows ARs of 1.0 (AA), 0.5 (AB), and 0 (BB). Loss of the distal sort arm loses the 0.5 values since only A or B alleles are present. In (**B**) the short arm of the chromosome and proximal long arm shows the typical diploid profile, but gain of the distal long arm generates AAB (AR = 0.67) and ABB (AR = 0.33) allelotypes which cause a distinct shift in the profile. Where the tumor is tetraploid (**C**), the ARs are normally 1.0, 0.75, 0.25, and 0.0 (see text) as seen for the short arm and proximal long arm of this chromosome. A single copy loss on a tetraploid background generates ARs of 0.67 and 0.33. In a balanced tetraploid situation (**D**) where the alleles are AAAA, AABB, or BBBB the ARs show a pattern similar to that seen in diploid samples (compare **A** and **D**) but where there is loss of a chromosome region on this background, as seen in the distal long arm in (**D**), the ARs become 0.67 and 0.33 as seen in triploid samples.

values deviate slightly from the expected ratios because of variation in individual SNP characteristics and background noise. When these ARs are plotted along the length of the chromosomes, a distinctive pattern is generated showing all three possibilities (**Fig. 5.5A**). Thus, where there is CNA_{loss}, or LOH on a diploid background, the only alternatives are A or B and this will result in a change in the profile as described for the proximal short arm shown in **Fig. 5.5A**. In general, ARs from long runs of consecutive homozygous loci will segregate into values of 1 and 0, regardless of copy number. In this case the ARs produced tend to bleed toward the middle of the AR profile, since the background signal has a greater impact on an overall signal generated by a single allele *(7)*.

In contrast, a region of single-copy gain (CN = 3) produces a distinctly different pattern of AR that results from potential allelic combinations of AAA (0), AAB (0.33), ABB (0.67), or BBB (1.0), as seen in **Fig. 5.5B** (unless all three copies derive from a single parental chromosome through some combination of loss and triplication). A two-copy gain (CN = 4) may produce either of two AR patterns, depending on whether the gain is due to a duplication of both parental chromosomes or due to triplication of one of the chromosomes. The latter (unbalanced) case of tetraploidy exhibits a pattern produced by four possible ARs: AAAA (0), AAAB (0.25), ABBB (0.75), or BBBB (1.0), as in the p-arm in **Fig. 5.5C**. On the other hand, a balanced tetraploid chromosome (**Fig. 5.5D**) produces a pattern of AR similar to that of a normal heterozygous diploid having possible allele combinations of AAAA (0), AABB (0.5), or BBBB (1.0). A loss of one copy on this tetraploid background, however, produces a shift in the ARs from 0.75 and 0.25 to 0.67 and 0.33, respectively, and is immediately obvious as shown in **Fig. 5.5D**.

The software tools to generate ARs as described above have only recently been developed and, at present, are only publicly available to support the Affymetrix 6.0 SMA through the Partek software suite. The data reported here were generated using a prototype of this tool, to analyze data from 100 K to 500 K SMAs.

4. Summary

The capability to define complex genetic events in tumor cells has evolved onto a single platform and while technically adequate to report these genetic changes, software development must progressively evolve to provide more meaningful interpretation. With the capability to define structural genetic changes at ultra high resolution in the tumor karyotype, the challenge is now to relate these events to gene expression changes in the same tumors. This can theoretically be achieved by overlaying the two datasets, although the challenges to achieve this computationally have been considerable. We have described one approach *(20)* to achieve this parallel analysis, and an alternative approach is discussed in **Chapter 15** of this volume.

References

1. Cowell, J. K., Barnett, G. and Nowak N. J. (2004) Characterization of the 1p/19q chromosomal loss in oligodendrogliomas using CGHa. *J. Neuropathol. Exp. Neurol.* **63**, 151–158.

2. Lo, K. C., Ma, C., Bundy, B. N., Pomeroy, S. L., Eberhart, C. G. and Cowell, J. K. (2007) Gain of 1q is a univariate negative prognostic marker for survival in medulloblastoma. *Clin. Cancer Res.* **13**, 7022–7028.

3. Cavenee, W. K., Dryja, T. P., Phillips, R. A., Benedict, W. F., Godbout, R., Gallie, B. L., Murphree, A. L., Strong, L. C. and White, R. L. (1983) Expression of recessive alleles by chromosomal mechanisms in retinoblastoma. *Nature* **305**, 779–784.

4. Cowell, J. K. and Hawthorn, L. (2007) The application of microarray technology to the analysis of the cancer genome. *Curr. Mol. Med.* **7**, 103–120.

5. Nannya, Y., Sanada, M., Nakazaki, K., Hosoya, N., Wang, L., Hangaishi, A., Kurokawa, M., Chiba, S., Bailey, D. K., Kennedy, G. C. and others. (2005) A robust algorithm for copy number detection using high-density oligonucleotide single nucleotide polymorphism genotyping arrays. *Cancer Res.* **65**, 6071–6079.

6. Lo, K. C., Bailey, D., Burkhardt, T., Gardina, P., Turpaz, Y. and Cowell, J. K. (2008) Comprehensive analysis of loss of heterozygosity events in glioblastoma using the 100 K SNP mapping arrays and comparison with copy number abnormalities defined by BAC array comparative genomic hybridization. *Genes Chromosomes Cancer* **47**, 221–237.

7. Gardina, P. J., Lo, K. C., Lee, W., Cowell, J. K. and Turpaz, Y. (2008) Ploidy status and copy number aberrations in primary glioblastomas defined by integrated analysis of allelic ratios, signal ratios and loss of heterozygosity on 500 K SNP mapping arrays. *BMC Genomics* In press.

8. Shankar, G., Rossi, M. R., McQuaid, D., Conroy, J. M., Gaile, D. G., Cowell, J. K., Nowak, N. J. and Liang, P. (2006) aCGH viewer: A generic visualization tool for aCGH data. *Cancer Inform.* **2**, 36–43.

9. Fridlyand, J., Snijders, A. M., Pinkel, D., Albertson, D. G. and Jain A. N. (2004) Hidden Markov models approach to the analysis of array CGH data. *J. Multivariate Anal.* **90**, 132–153.

10. Lai, W. R., Johnson, M. D., Kucherlapati, R. and Park, P. J. (2005) Comparative analysis of algorithms for identifying amplifications and deletions in array CGH data. *Bioinformatics* **21**, 3763–3770.

11. Mitelman, F., Johansson, B. and Mertens, F. (Eds.) (2008). Mitelman Database of Chromosome Aberrations in Cancer. http://cgap.nci.nih.gov/Chromosomes/Mitelman

12. Rossi, M. R., Gaile, D., LaDuca, J., Matsui, S. I, Conroy, J., McQuaid, D., Chervinsky, D., Eddy, R., Chen, H-S., Barnett, G., Nowak, N. J. and Cowell, J. K. (2005) Identification of consistent novel megabase deletions in low-grade oligodendrogliomas using array-based comparative genomic hybridization. *Genes Chromosomes Cancer* **44**, 85–96.

13. Lo, K. C., Rossi, M. R., LaDuca, J., Hicks, D. G., Turpaz, Y. and Hawthorn, L. (2007) Candidate gliobastoma development gene identification using concordance between copy number abnormalities and gene expression level changes. *Genes Chromosomes Cancer* **46**, 875–894.

14. Cowell, J. K. (1982) Double minutes and homogeneously staining regions: Gene amplification in mammalian cells. *Ann. Rev. Genet.* **16**, 21–59.

15. Stark, G. R. (1993) Regulation and mechanisms of mammalian gene amplification. *Adv. Cancer Res.* **61**, 87–113.

16. Lockwood, W. W., Chari, R., Chi, B. and Lam, W. L. (2006) Recent advances in array comparative genomic hybridization technologies and their applications in human genetics. *Eur. Hum. Genet.* **14**, 139–148.

17. Knudson, A. G. (1971) Mutation and cancer: Statistical study of retinoblastoma. *Proc. Natl. Acad. Sci. USA* **68**, 820–823.

18. Cowell, J. K., LaDuca, J., Rossi, M. R., Burkhardt, T., Nowak, N. J. and Matsui, S-I. (2005) Molecular characterization of the t(3;9) translocation associated with immortalization in the MCF10A cell line. *Cancer Genet. Cytogenet.* **163**, 23–29.

19. Rossi, M. R., LaDuca, J., Matsui, S-I., Nowak, N. J., Hawthorn, L. and Cowell, J. K. (2005) Novel amplicons on the short arm of chromosome 7 identified using high resolution array CGH contain over expressed genes in addition to EGFR in glioblastoma multiforme. *Genes Chromosomes Cancer* **44**, 392–404.

20. Lo, K. C., Shankar, G., Turpaz, Y., Bailey, D., Rossi, M. R., Burkhardt, T., Liang, P and Cowell, J. K. (2007) Overlay tool for aCGHviewer: An analysis module built for aCGHViewer used to combine different microarray platforms for visualization. *Cancer Inform.* **3**, 307–319.

Chapter 6

Molecular Inversion Probe Assay for Allelic Quantitation

Hanlee Ji and Katrina Welch

Abstract

Molecular inversion probe (MIP) technology has been demonstrated to be a robust platform for large-scale dual genotyping and copy number analysis. Applications in human genomic and genetic studies include the possibility of running dual germline genotyping and combined copy number variation ascertainment. MIPs analyze large numbers of specific genetic target sequences in parallel, relying on interrogation of a barcode tag, rather than direct hybridization of genomic DNA to an array. The MIP approach does not replace, but is complementary to many of the copy number technologies being performed today. Some specific advantages of MIP technology include: less DNA required (37 ng vs. 250 ng), DNA quality less important, more dynamic range (amplifications detected up to copy number 60), allele-specific information "cleaner" (less SNP cross-talk/contamination), and quality of markers better (fewer individual MIPs versus SNPs needed to identify copy number changes). MIPs can be considered a candidate gene (targeted whole genome) approach and can find specific areas of interest that otherwise may be missed with other methods.

Key words: Molecular inversion probe, genotyping, gene copy number, microarrays, single-nucleotide polymorphisms, alleles.

1. Introduction

A molecular inversion probe (MIP) is a single oligonucleotide that recognizes and hybridizes to a specific genomic target sequence with two inverted recognition complementary flanks ranging from 20 to 30 nucleotides (1). The total length of the MIP is 120 nucleotides. After the probe hybridizes to the target DNA, a single base pair gap exists in the middle of the two recognition sequences (**Fig. 6.1**). The gap can represent either a single-nucleotide polymorphism (SNP) or a non-polymorphic nucleotide. With the addition of the specific nucleotide to fill the gap and subsequent

Jonathan R. Pollack (ed.), *Microarray Analysis of the Physical Genome: Methods and Protocols, vol. 556*
© Humana Press, a part of Springer Science+Business Media, LLC 2009
DOI 10.1007/978-1-60327-192-9_6 Springerprotocols.com

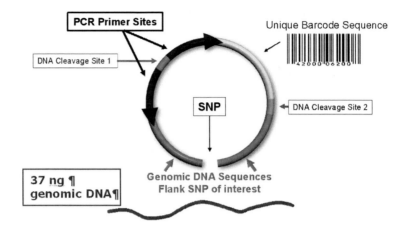

Fig. 6.1. Molecular inversion probe (MIP). Each MIP is 120 bp oligonucleotide, with a unique gap fill for SNP of interest. Each probe contains unique tag (barcode) sequence corresponding to interrogated SNP.

ligation after the probe specifically anneals to its complementary genomic sequence, the probe undergoes an intramolecular rearrangement that allows the amplification of a barcode sequence unique to each oligonucleotide. The barcode is queried via a microarray and intensity of a specific barcode reflects a specific SNP and is also quantitative in terms of copy number (2–4).

The flow process of the reaction is as follow:

1. In a single sample volume, the MIP probes are mixed with genomic DNA.

2. Subsequently annealing is carried out. A "gap fill" ligation reaction is then carried out with a dNTP in the tube.

3. The samples are then amplified using a single set of common primers and the fragment content of each reaction is hybridized to a standard oligonucleotide chip array.

4. After the hybridization on the DNA chip, the components are decoded. The relative base incorporation is measured by the fluorescence signal intensity at the corresponding complementary barcode tag site on the DNA array.

The advantages of MIP technology are derived from the intramolecular nature of the assay (1–4). A single molecule (probe) hybridizes to genomic DNA with two recognition sequences. Even though there are two hybridization events, the second is extremely rapid since the effective local concentration of the second site is in the micromolar range. For this reason, much less probe is required as compared to other molecule assays. We typically use between 5 and 20 attomoles (10^{-18} mol) of each probe in a 1,000 multiplex reaction for 5–20 femtomoles (10^{-15} mol) of total probe. The benefits of this are that more probes can be added into a single reaction (higher multiplexing) and lower concentration of each probe reduces the likelihood of two probes interacting with each other (lower backgrounds). In the initial report, this technology was used for standard SNP genotyping at high levels of multiplexing *(1)*. We found that this approach was highly quantitative and highly accurate through multiple hybridization and enzymatic processing events *(2–4)*.

There are a number of technologies that assess copy number on the whole-genome scale such as array comparative genomic hybridization (CGH). Formats include using bacterial artificial chromosome (BAC) CGH, spotted cDNAs, and more recently several types of oligonucleotides arrays including ones which rely on SNP sequences. Some of the newer CGH methodologies allow for allelic information to be obtained. The utility of measurement of allele copy number includes the identification of loss of heterozygosity (LOH) events and the allelic genotype at amplified loci. Likewise, MIP technology has been demonstrated to be a robust platform for human genomics application such as germline genotyping and copy number variation (CNV) in cancer *(2)*. This system has a number of strengths including minimal DNA requirements, more flexible sequence design constraints because of its reliance on a barcode intermediate as opposed to direct array hybridization, and ability to use a ubiquitous barcode array that is inexpensive (<$90 per array). Because a barcode intermediate is used instead of direct genomic DNA hybridization to an array, one can query any unique, nonrepetitive sequence for simultaneous genotyping and copy number changes (e.g. allelic quantitation) without having to optimize for specific hybridization parameters of array CGH or oligonucleotide arrays. Expanding the number of loci is simply a matter of producing more probes. Gene copy determination is sequence specific, meaning that the resolution of determining gene copy number alterations is oftentimes at an individual sequence level for high-performing probes. This enables extremely high-resolution delineation of copy number changes which are specific to a given gene. As already mentioned, less DNA is required (37 ng), and the source of the DNA and its quality is less important to accurate allelic quantitation with sources such as paraffin-embedded tissues *(3, 4)*. The current number of loci which can be analyzed ranges in the tens of thousands but there are efforts to expand probe sets into the hundreds of thousands.

2. Materials

2.1. Reagents and Arrays

1. Molecular inversion probe pool (Affymetrix, Santa Clara, CA)

2. ExoMix – Affymetrix GeneChip® Custom SNP Kit (Affymetrix, Santa Clara, CA)

3. 10X Buffer A – Affymetrix GeneChip® Custom SNP Kit (Affymetrix, Santa Clara, CA)

4. Gap Fill Mix – Affymetrix GeneChip® Custom SNP Kit (Affymetrix, Santa Clara, CA)

5. AmpMix – Affymetrix GeneChip® Custom SNP Kit (Affymetrix, Santa Clara, CA)

6. Second-stage PCR Mix – Affymetrix GeneChip® Custom SNP Kit (Affymetrix, Santa Clara, CA)

7. Cleavage Enzyme – Affymetrix GeneChip® Custom SNP Kit (Affymetrix, Santa Clara, CA)

8. Cleavage Tube – Affymetrix GeneChip® Custom SNP Kit (Affymetrix, Santa Clara, CA)

9. HY Digest Mix – Affymetrix GeneChip® Custom SNP Kit (Affymetrix, Santa Clara, CA)

10. Stain Cocktail – Affymetrix GeneChip® Custom SNP Kit (Affymetrix, Santa Clara, CA)

11. Hyb Cocktail – Affymetrix GeneChip® Custom SNP Kit (Affymetrix, Santa Clara, CA)

12. dNTPs – Affymetrix GeneChip® Custom SNP Kit (Affymetrix, Santa Clara, CA)

13. Arrays, Universal Tag 70 K – Affymetrix GeneChip® Custom SNP Kit (Affymetrix, Santa Clara, CA)

14. Wash A – Affymetrix GeneChip® Custom SNP Kit (Affymetrix, Santa Clara, CA)

15. Wash B – Affymetrix GeneChip® Custom SNP Kit (Affymetrix, Santa Clara, CA)

16. PicoGreen dsDNA Quantitation Reagent (Molecular Probes, Inc., Eugene, OR)

17. TE (pH 8.0) buffer

18. Stratagene Taq DNA Polymerase (Stratagene, San Diego, CA)

19. TITANIUM™ Taq Polymerase (Clontech, Mountain View, CA)

20. 3% Precast Gel, Ready Agarose 96 Plus, TBE (BioRad, Hercules, CA)

21. 1 kb Molecular Ruler (BioRad, Hercules, CA)

22. BSA (10 mg/mL)

23. 10X TBE Buffer

24. SAPE (1 mg/mL)

2.2. Disposables

1. 200 μL PCR tubes in strips of 12

2. BD Falcon microplate

3. Filtered pipette tips

4. Kim Wipes

5. Disposable gloves

6. 50 mL Reagent Reservoir

7. 1.5 mL Eppendorf tube

8. 96-well plates

9. 384-well plates

10. Thermo-Fast® 96 Rigid Skirted PCR Plate, opaque black (Thermo Fisher Scientific, Rockford, IL)

11. Clear Adhesive Films (Applied Biosystems, Foster City, CA)

2.3. Large Equipment

1. Perkins Elmer Victor 2 1420 Multilabel Counter (Perkin Elmer, Waltham, MA)

2. Two refrigerators, 4°C refrigerator; 6 cu ft.

3. Two freezers, −20°C; deep freeze; manual defrost; 17 cu ft.[2]

4. Thermal cycler, e.g., GeneAmp® PCR System 9700 Thermal Cycler (Applied Biosciences, Foster City, CA)

5. Two tabletop centrifuges with swinging bucket rotor holding 96-well plates (e.g., Sorvall Legend RT tabletop centrifuge – Thermo Fisher Scientific Inc., Waltham, MA)

6. Eppendorf® Multipurpose Centrifuge 5804 2 VWR Intl. 53513-800 1 Pre-Amp Lab

7. GeneChip® Hybridization Oven 640 with carriers (Affymetrix, Santa Clara, CA)

8. GeneChip® Fluidics Station 450 (Affymetrix, Santa Clara, CA)

9. Affymetrix GeneChip® Scanner (Affymetrix, Santa Clara, CA)

10. Standard UV Gel Imager

2.4. Small Equipment

1. Two sets of single channel pipetters (e.g., Ranin P-2, P-10, P-20, P-200, P-1000 – Ranin, Oakland, CA)

2. Two sets of 12-channel pipetters (Ranin P-10, P-20, P-200 – Ranin, Oakland, CA)

3. Two sets of 24-channel pipetters (Ranin P-10, P-20, P-200 – Ranin, Oakland, CA)

4. Eppendorf tube rack

5. Two Galaxy Mini Centrifuges or similar model (interchangeable for microtubes and strip tubes)

6. Two vortexers

7. Gel Box, wide Mini-Sub Cell GT cell (BioRad, Hercules, CA)

8. Portable pipette aid

2.5. Computers

1. 7G Instrument Control Workstation: The Instrument Control Workstation is installed in the Post-Amp Lab. It controls the scanner and the fluidics station. This workstation will use Microsoft® Windows XP operating system with Service Pack 1 or 2. This workstation should be operating the Affymetrix GeneChip® Operating Software (GCOS). It should be appropriately networked as per the installation instructions.

2. MIP Post-Amp Lab Workstation: The MIP Post-Amp Lab Workstation includes Microsoft Windows XP operating system with Service Pack 2. It should be appropriately networked as per the installation instructions.

    a. Microsoft® SQL Server 2000, personal edition

    b. Java 2 Platform, Standard Edition, release 1.4.2

    c. Apache Tomcat Web Server 4.9.1

    d. GeneChip® Targeted Genotyping Analysis Software (GTGS) (Java-based)

    e. Microsoft Office Professional

    f. GCOS Runtime Libraries

3. Methods

This method describes the analysis of 48 samples. On previous experience, we have found that data quality is better using larger sample batches. Therefore, we typically accrue enough samples to run a minimum of 48 per any given experimental run. Each sample will be tagged with two dinucleotide mixes, dATP/dTTP and dCTP/dGTP, and ultimately each sample will be analyzed by two microarrays totaling 96 microarrays. The two microarrays per sample will correspond to the two dinucleotide mixes each samples will be tagged with.

We recommend running 5–10 normal diploid genome samples simultaneously that were processed for genomic DNA in a similar fashion to tumor samples. This is necessary for downstream data processing.

Each work area should be separated by a physical boundary such as a wall; it is ideal to have two rooms. The two work areas are namely the pre-amplification lab (pre-amp) and post-amplification lab (post-amp). Work flow should always proceed forward from pre-amp to post-amp.

3.1. Measuring the Concentration of Genomic DNA (Pre-amp)

1. Prepare a working stock of 1X TE Buffer (pH 8.0) by diluting a concentrated stock using sterile DNAse-free water.

2. Prepare lambda DNA standards by making a dilution series of the 100 ng/μL stock solution with the following concentrations: 0, 2, 8, 14, and 20 ng/μL. Dilute the stock using 1X TE (pH 8.0) buffer and bring the final volume to 1 mL (*see* **Table 6.1**).

Table 6.1
Concentration of lambda DNA standards in both the first dilution series and in the final concentration in the assay

Volume (μL) of 100 μg/mL DNA stock	Volume (μL) 1X TE buffer	Final DNA concentration (μg/mL)	DNA concentration in assay (μg/mL)
200	800	20	1
140	986	14	0.7
80	982	8	0.4
20	998	2	0.1
0	1,000	Blank	Blank

3. Add 10 μL of each lambda DNA standard to a well of a Thermo-Fast® 96 Rigid Skirted PCR Plate. Once the standard DNA solutions have been aliquoted, store at 4°C for future use.

4. To each well containing 10 μL of lambda DNA standard, add 90 μL 1X TE (pH 8.0). Record the location of each sample. Set the plate aside on the bench.

5. Thaw all DNA samples which have an unknown concentration. Vortex all samples repeatedly to ensure the samples are homogeneous throughout. Pulse centrifuge to remove excess liquid from the side of the tube.

6. Dilute the DNA samples by adding 2 μL DNA to 198 μL 1X TE (pH 8.0) to a fresh 1.5 mL Eppendorf tube. Vortex to mix.

7. Add 10 μL of this dilute DNA and 90 μL 1X TE (pH 8.0) to an assigned well: the 96-well plate which the lambda DNA standards were previously aliquoted onto. The total dilution of the DNA samples is now 1,000-fold the original stock. Complete this for each sample.

8. In a 1.5 mL Eppendorf tube prepare a working stock of PicoGreen reagent by diluting the PicoGreen 200-fold in 1X TE (pH 8.0). For example, pipette 5 μL PicoGreen in 995 μL TE which is enough for exactly 10 samples. PicoGreen reagent is degraded by light; all original and working stocks should therefore be wrapped in aluminum foil to prevent light exposure.

9. Add 100 μL dilute PicoGreen reagent to the wells containing DNA and lambda DNA standards – the dilution of the sample DNA is now 2,000-fold. Use a 12-channel pipette when processing large numbers of samples to avoid time lapse between incubations.

10. Allow the PicoGreen reaction to incubate for 2–5 min at room temperature in the dark.

11. Measure fluorescence using a fluorescent scanner with a microplate reader such as the Perkin Elmer fluorimeter.

12. Plot the standard curve intensity data against the known DNA concentration of the standards and derive a linear regression. Use this equation to determine the concentration of each individual sample.

3.2. MIP Annealing – Preparation of the DNA Samples and 384-Well Annealing Plate (Pre-amp)

1. Based on the quantitation of the DNA from the PicoGreen assay, normalize each genomic DNA sample to 8 ng/μL. Pipette 10–20 μL of each sample into rows A, B, C, and D of a 96-well plate and record the well location for each sample. Label this plate *Sample Plate* and place the plate on ice. All remaining steps will refer to this plate as the *Sample Plate*.

2. Label the 384-well plate with the title *Anneal Plate*. This 384-well plate will be referenced as the *Anneal Plate* in all subsequent steps. Place the plate on ice.

3. In a 1.5 mL Eppendorf tube, prepare the *Probe Master Mix*. This master mix contains 10X Buffer A, Probe, and Enzyme A. To the 1.5 mL Eppendorf tube add 45 μL 10X Buffer A, 66.6 μL Probe, and 2.7 μL Enzyme A. The master mix is designed to create enough volume for 48 samples plus an additional 10% leaving room for pipette error. Enzyme A should be added last as it is extremely heat sensitive; minimize warming of this enzyme by keeping it on ice at all times.

4. Aliquot out 9.2 μL of the *Probe Master Mix* described above (10X Buffer A + Probe + Enzyme A) into every other well in row A up to well A23 on the *Anneal Plate*. More explicitly wells A1, A3, A5, A7, A9, A11, A13, A15, A17, A19, A21, and A23 on the *Anneal Plate* should contain *Probe Master Mix*. This row containing the master mix is used the same way 12-strip PCR tubes would be used; it is simply an easier way to make additional aliquots to subsequent rows in the same plate. The master mix is placed in every other lane to allow the use of a 12-channel pipette in the steps which follow.

5. Using a 12-channel pipette dispense 1.9 μL/well of the *Probe Master Mix* into rows B, C, D, and E of the *Anneal Plate*. The 12-channel pipette will dispense the *Probe Master Mix* into every other column on the *Anneal Plate* similar to the way the *Probe Master Mix* was dispensed into row A. The 1.9 μL will be taken from the 9.2 μL which was aliquoted into the 12 wells in row A (Step 3).

6. From the 96-well *Sample Plate*, using a 12-channel pipette, add 4.7 μL of the genomic DNA into the wells containing the *Probe Master Mix,* mix up and down 10 times to mix while avoiding bubbles. More explicitly, 4.7 μL of sample from row A, row B, row C, and row D on the *Sample Plate* should be transferred to row B, row C, row D, and row E on the *Anneal Plate*, respectively. Because the DNA was transferred using a 12-channel pipette, the sample will be distributed in every other well just as the *Probe Master Mix* was dispensed. Ensure that the well location of each DNA sample is recorded.

7. Tightly seal the *Anneal Plate* with a clear adhesive film.

8. Spin the *Anneal Plate* in a balanced tabletop centrifuge at 1,000 rpm for 15 s. All subsequent spins should be done in this tabletop centrifuge.

9. Transfer the *Anneal Plate* to the thermal cycler and start the *Meg Anneal program* (*see* **Fig. 6.2a**).

10. One minute into the 58°C incubation pause the thermal cycling program, take the *Anneal Plate* out and place it on ice for 2 min.

11. Spin the *Anneal Plate* at 1,000 rpm for 15 s. Put the *Anneal Plate* back onto the thermal cycler and resume the thermal cycling program, 58°C anneal overnight. This step is designed to remove liquid from sides of the wells which may have formed from bubbles bursting during the 95°C stage. It is imperative that the entire volume of the reaction takes place in the overnight annealing.

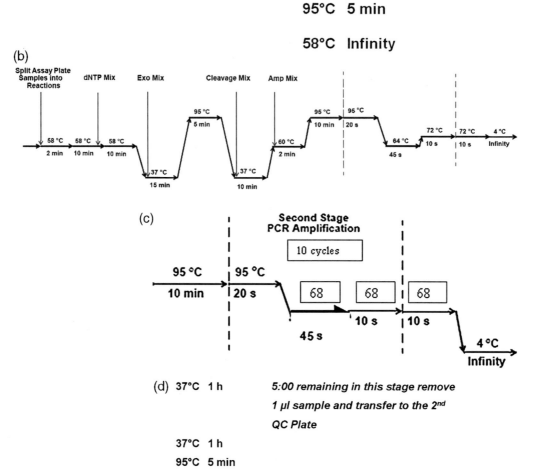

Fig. 6.2. Thermal cycling programs. Each program should be programmed into the appropriate thermal cycler, pre-amp lab, or post-amp lab, prior to commencement of the MIP assay.
(a) *Meg Anneal* thermal cycling program to be programmed in the pre-amp lab.
(b) *Meg 3-5-7 K* thermal cycling program to be programmed in the pre-amp lab.

58°C	2 min	
58°C	10 min	*1:00 remaining in this stage add the dNTPs*
58°C	10 min	
37°C	15 min	*14:00 remaining in this stage add the Exo Mix*
95°C	5 min	
37°C	10 min	*9:00 remaining in this stage add the Cleavage Mix*
60°C	2 min	*At this stage add the Amp Mix*
<u>95°C</u>	<u>10 min</u>	
95°C	20 s	
64°C	45 s	22 cycles
<u>72°C</u>	<u>10 s</u>	
72°C	10 s	
4°C	Infinity	

12. Annealing time is 16 h (+/− 1 h), e.g., start annealing at 5 p.m. on day 1. The next stage can start at 9 a.m. on day 2. Best results are obtained when the annealing time is kept consistent between experiments. For example, if the annealing time is 16 h and 30 min for one experiment try to replicate that time for the experiment which follows.

3.3. MIP First-Stage PCR – Creating the Assay Plate, Aliquoting Gap Fill and dNTPs (Pre-amp)

1. On day 2, take a new 96-well plate and label it *Assay Plate*. Place the plate on ice. The remaining steps will refer to this 96-well plate as the *Assay Plate*.

2. Into 12-strip PCR tubes, pipette 11 μL of the gap fill mix into all 12 wells of the strip. The gap fill mix is a pre-made reagent. Label the strip tube *Gap Fill* and set the 12-strip PCR tubes aside on ice.

3. Prepare *1X Buffer A* by adding 100 μL of the 10X Buffer A solution into 900 μL of H_2O. The 10X Buffer A solution is a pre-made reagent.

4. Into 12-strip PCR tubes, pipette 80 μL of *1X Buffer A* into all 12 wells of the strip. Label the strip tube as *1X Buffer A* and set aside on ice.

5. Remove the *Anneal Plate* from the thermal cycler where it has been held at 58°C overnight and place the plate on ice for 2 min. Spin the plate at 1,000 rpm for 15 s.

6. Using a 12-channel pipette and the PCR strip tubes containing the 11 μL of gap fill mix previously aliquoted as your stock, add 1.25 μL of the gap fill mix to each well containing sample on the *Anneal Plate*. The wells containing sample are every other column in rows B, C, D, and E up to column number 23. Mix well by pipetting up and down three times avoiding bubbles.

7. Using a 12-channel pipette and the PCR strip tubes containing the 80 μL of *1X Buffer A* previously aliquoted as your stock, add 12 μL of the *1X Buffer A* to each well containing

(c) Hy-Titanium-10 cycles *thermal cycling program to be programmed in the post-amp lab.*

95°C   10 min
95°C   20 s
68°C   45 s   22 cycles
68°C   10 s
68°C   10 s
4°C   Infinity

(d) Meg Hydigest – *a thermal cycling program to be programmed in the post-amp lab.*

sample on the *Anneal Plate*. These are the same wells the gap fill mix was just added to. Mix well by pipetting up and down 20 times avoiding bubbles.

8. All samples from the *Anneal Plate* will now be transferred to the *Assay Plate*. Using a 12-channel pipette transfer 9 μL from row B of the 384-well annealing plate to rows A and B of the *Assay Plate*. Repeat this step transferring this time 9 μL from row C of the *Anneal Plate* to rows C and D of the *Assay Plate*. Now transfer 9 μL from row D of the *Anneal Plate* to rows E and F of the *Assay Plate*. Finally transfer 9 μL from row E of the *Anneal Plate* to rows G and H of the *Assay Plate*. Seal the wells of the *Assay Plate* with a clear adhesive film. Spin the *Assay Plate* at 1,000 rpm for 15 s.

9. At this point there should be very little or nothing left in the *Anneal Plate*.

10. Start the thermal cycler program *Meg 3-5-7 K* (*see* **Fig. 6.2b**). Once the thermal cycler has reached 58°C, load the *Assay Plate* and close the lid. The reagents which will be prepared in the following steps will be added to the *Assay Plate* while it remains on the thermal cycler. Be aware of the thermal cycler and which stage it is at as dNTPs addition will occur when the timer reads 1 min remaining in the second 58°C period.

11. The nucleotides, dNTPs, come in a 96-well plate. Spin the dNTP plate at 1,000 rpm for 15 s. Place the plate on ice.

12. Carefully remove the adhesive film covering the wells. The rows on the dNTP plate will be clearly labeled indicating which nucleotide is in each row.

13. The dNTP plate contains approximately 20 μL of nucleotide per well. Using a multichannel pipette prepare a dinucleotide mix by combining 20 μL dATP with 20 μL dTTP to a labeled 12-strip PCR tubes. Repeat this creating a dinucleotide mix of dCTP with dGTP and combining 20 μL of each of these two nucleotides to a new labeled 12-strip PCR tubes. Each 12-strip PCR tubes, dATP/dTTP and dCTP/dGTP, will now be containing approximately 40 μL per well.

14. When the thermal cycler timer reads 1 min remaining for the second 58°C period, pause the thermal cycler, remove the *Assay Plate*, and place it on ice for 1 min.

15. Carefully remove the clear adhesive film from the *Assay Plate*.

16. Using a 12-channel pipette, add 4 μL of the dATP/dTTP dinucleotide mix to rows A, C, E, and G of the *Assay Plate* – this is every other row. Mix will by pipetting up and down 20 times avoiding bubbles.

17. Using a 12-channel pipette, add 4 µL of the dCTP/dGTP dinucleotide mix to rows B, D, F, and H of the *Assay Plate* – again this is every other row. Mix well by pipetting up and down 20 times avoiding bubbles.

18. Seal the *Assay Plate* with a clear adhesive film. Vortex the plate to mix. Spin the plate at 1,000 rpm for 15 s.

19. Return the *Assay Plate* to the thermal cycler, close the lid, and resume the *Meg 3-5-7 K* program. Be conscious of the thermal cycling program. New reagents will be added to the plate with 14 min remaining in the first 37°C period. These reagents will be prepared in *MIP First-Stage PCR – aliquoting ExoMix, AmpMix, and Cleavage Reagent Mix.*

3.4. MIP First-Stage PCR – Aliquoting ExoMix, AmpMix, and Cleavage Reagent Mix (Pre-amp)

1. Prepare two 12-strip PCR tubes. Pipette 45 µL of the exonuclease mix into each tube of one of the 12-strip PCR tubes. The exonuclease mix is a pre-made reagent. Using a 12-channel pipette transfer 22.5 µL of the exonuclease mix from the first 12-strip PCR tubes into the second 12-strip PCR tubes. Label the two 12-strip PCR tubes *Exo Mix* and set aside on ice.

2. When the thermal cycler reads 14:00 min remaining in the first 37°C period, pause the *Meg 3-5-7 K* program.

3. Remove the *Assay Plate* from the thermal cycler and place on ice for 2 min.

4. Spin the *Assay Plate* at 1,000 rpm for 15 s and return the plate to ice.

5. Carefully remove the clear adhesive film from the *Assay Plate*.

6. Using a 24-channel pipette, transfer 4 µL of *Exo Mix* from the two sets of 12-strip PCR tubes into each well on the *Assay Plate* mixing 20 times with each addition. Avoid bubbles.

7. Reseal the *Assay Plate* with clear adhesive film and spin the plate at 1,000 rpm for 15 s.

8. Place the *Assay Plate* back on the thermal cycler, close the lid, and resume the *Meg 3-5-7 K* thermal cycling program.

9. Prepare the *Cleavage Reagent* by adding 24 µL of the pre-made cleavage enzyme to each 3 mL cleavage tube, also pre-made. This mixture of cleavage reagent and cleavage enzyme will now be referred to as the *Cleavage Reagent Mix*. Place the mix on ice.

10. Prepare two 12-strip PCR tubes. Pipette 240 µL of the *Cleavage Reagent Mix* into each tube of one of the two 12-strip PCR tubes. Using a 12-channel pipette transfer 120 µL of the

Cleavage Reagent Mix from the first 12-strip PCR tubes into the second 12-strip PCR tubes. Label the 12-strip PCR tubes *Cleavage Mix* and place on ice.

11. Prepare the *Amp Mix* by adding 67 µL Stratagene Taq Polymerase to the pre-made amp mix tube. This mixture of Stratagene Taq Polymerase and amp mix will now be referred to as the *Amp Mix*. Place the mix on ice.

12. Prepare two 12-strip PCR tubes. Pipette 240 µL of the *Amp Mix* into each tube of one of the 12-strip PCR tubes. Then using a 12-channel pipette transfer 120 µL of the *Amp Mix* from the first 12-strip PCR tubes into the second 12-strip PCR tubes. Label the 12-strip PCR tubes *Amp Mix* and set aside on ice.

13. When the thermal cycler reads 9:00 min remaining in the second 37°C period, pause the *Meg 3-5-7 K* program.

14. Leaving the *Assay Plate* on the thermal cycler, carefully remove the clear adhesive film.

15. Using a 24-channel pipette, transfer 25 µL of *Cleavage Mix* from the two sets of 12-strip PCR tubes previously made (Step 10) into each well on the *Assay Plate*. Mix 10 times by pipetting up and down with each addition of *Cleavage Mix*. Avoid bubbles.

16. Reseal the *Assay Plate* with clear adhesive film, close the lid, and resume the *Meg 3-5-7 K* thermal cycling program.

17. When the thermal cycler reaches the first 60°C period, pause the *Meg 3-5-7 K* program.

18. Leaving the *Assay Plate* on the thermal cycler; carefully remove the clear adhesive film.

19. Using a 24-channel pipette, transfer 25 µL of *Amp Mix* from the two sets of 12-strip PCR tubes previously made (Steps 11 and 12) into each well on the *Assay Plate*. Mix 10 times by pipetting up and down with each addition of *Amp Mix*. Avoid bubbles.

20. Reseal the *Assay Plate* with clear adhesive film, close the lid, and resume the *Meg 3-5-7 K* thermal cycling program.

21. Once the *Meg 3-5-7 K* thermal cycler program has reached completion all subsequent steps should be run in the post-amp lab. No products from this point should be moved into the pre-amp lab.

3.5. MIP Quality Control – Checking the First-Stage PCR (Post-amp)

1. Once the *Meg 3-5-7 K* program has reached completion, remove the *Anneal Plate* from the thermal cycler, transfer the *Anneal Plate* to the post-amp lab, and place the plate on ice. Carefully remove the clear adhesive film.

2. Label a new 96-well plate *First QC Plate* and place the plate on ice. Using a 24-channel pipette transfer 15 μL from each well on the *Anneal Plate* into the corresponding wells on the *First QC Plate*.

3. To each well on the *First QC Plate* add 2 μL loading dye to the sample.

4. Prepare a working stock of 1X TBE Buffer. For instance, add 100 mL 10X TBE Buffer to 900 mL H_2O.

5. Load the BioRad precast 3% gel into a gel box containing 1X TBE Buffer. Ensure the empty wells of the gel are completely submerged with buffer.

6. In the first and last well of the agarose gel, load 2.5 μL BioRad 1 kb Ladder.

7. Using a 12-channel pipette load each row of the *First QC Plate* into the wells on the gel.

8. Run the gel for 13 min at 150 V.

9. Visualize the gel using a standard UV gel imager. A clear distinct band should be present at 120 bp for each sample. No bands should appear below or above the prominent band.

3.6. MIPs Second-Stage PCR – Aliquoting Allele Tube Mix and First-Stage PCR to the Label Plate (Post-amp)

1. Prepare the *Allele Tube Mix* by adding 22 μL of Titanium Taq Polymerase to each Second-Stage PCR Mix tube, these are both pre-made reagents. Mix by pipetting up and down at least 10 times.

2. Pour the *Allele Tube Mix* into a 50 mL reagent reservoir and place on ice.

3. Take a new 96-well plate and label it *Label Plate*. This plate will be referred to the *Label Plate* in all subsequent steps. Place the plate on ice.

4. Using a 12-channel pipette, aliquot 15.5 μL of the *Allele Tube Mix* from the 50 mL reagent reservoir into all wells on the *Label Plate*.

5. Prepare the *Assay Plate* by removing the clear adhesive film and placing the plate on ice.

6. Using a 12-channel pipette, transfer 2 μL from each row of the *Assay Plate* to the corresponding row on the *Label Plate*. The *Label Plate* will also contain 15.5 μL of the *Allele Tube Mix* which was aliquoted in Step 4.

7. Set a 12-channel pipette to dispense 10 μL. Using this setting, mix the contents of the *Label Plate* in each row by pipetting up and down five times avoiding bubbles.

8. Seal the *Label Plate* with clear adhesive film and spin the plate at 1,000 rmp for 15 s.

9. Place the *Label Plate* on a thermal cycler and close the lid.

10. Start the *HY-Titanium-10 cycle* program (*see* **Fig. 6.2c**).

11. When the *HY-Titanium-10 cycle* program is complete, remove the *Label Plate* from the thermal cycler and spin at 1,000 rpm for 15 s. Return the plate to ice.

3.7. MIP Target Digest – Creating the Hyb Plate and Aliquoting the Digest Mix (Post-amp)

1. Prepare a new 96-well plate by labeling it *Hyb Plate*. All subsequent steps will refer to this plate as the *Hyb Plate*.

2. Carefully remove the clear adhesive film from the *Label Plate*.

3. Using a 12-channel pipette, transfer 17 μL from each well of the *Label Plate* to the corresponding well on the *Hyb Plate*.

4. Prepare two 12-strip PCR tubes. Pipette 6 μL of the HY Digest Mix into each tube of one of the 12-strip PCR tubes. Using a 12-channel pipette transfer 3 μL of the HY Digest Mix from the first 12-strip PCR tubes into the second 12-strip PCR tubes. Label the 12-strip PCR tubes *HY Digest Mix*.

5. Using a 24-channel pipette, transfer 1.6 μL of *HY Digest Mix* from the two sets of 12-strip PCR tubes previously made, into each well on the *Hyb Plate*. The *Hyb Plate* will also contain 17 μL of sample transferred in Step 2. Set a 24-channel pipette to 15 μL and mix each well by pipetting up and down 20 times avoiding bubbles.

6. Seal the *Hyb Plate* with clear adhesive film and spin the plate at 1,000 rpm for 15 s.

7. Plate the *Hyb Plate* on the thermal cycler and start the *Meg Hydigest-a* program (*see* **Fig. 6.2d**).

3.8. MIP Quality Control – Checking the Second-Stage PCR (Post-amp)

1. Prepare a new 96-well plate and label it *Second QC Plate*.

2. When the thermal cycler reads 5 min remaining in the second 37°C period of the *Meg Hydigest-a* program, pause the thermal cycler and remove the *Hyb Plate*. Place the plate on the bench.

3. Remove the clear adhesive film from the *Hyb Plate* and using a 24-channel pipette remove 1 μL from each well and transfer it to the corresponding well of the *Second QC Plate*.

4. Reseal the *Hyb Plate* with clear adhesive film, return the plate to the thermal cycler, and resume the *Meg Hydigest-a* program.

5. To the *Second QC Plate* add 7 μL H_2O to each well. This can be done by adding H_2O to a 50 mL reagent reservoir and using a 24-channel pipette to aliquot the H_2O to the wells on the *Second QC Plate*.

6. Add 2 μL loading dye to each well of the *Second QC Plate*.

7. Load a BioRad precast 3% *gel* into a gel box containing *1X TBE Buffer*. Ensure the empty wells of the gel are completely covered with buffer.

8. In the first and last well of the agarose gel load 2.5 μL BioRad 1 kb Ladder.

9. Using a 12-channel pipette load each row of the *Second QC Plate* into the wells on the gel.

10. Run the gel for 13 min at 150 V.

11. Visualize the gel using a standard UV gel imager. A clear distinct band should be present at 80 bp. No bands should appear below or above the prominent band.

3.9. MIP Hybridization – Creating the Digest Plate and Aliquoting the Hyb Cocktail (Post-amp)

1. When the *Meg Hydigest-a* program has reached completion remove the *Hyb Plate* and place it on ice.

2. Prepare a new 96-well plate and label the plate *Digest Plate*. All subsequent steps will refer to this plate as the *Digest Plate*. Place this plate on ice.

3. Using a 12-channel pipette, transfer 1 μL of material from all wells on the *Hyb Plate* to the corresponding wells on the *Digest Plate*.

4. Prepare 12-strip PCR tubes. Pipette 294 μL of the Hyb Cocktail, a pre-made reagent, into each tube of the 12-strip PCR tubes. Label the 12-strip PCR tubes *Hyb Cocktail*. Place the 12-strip PCR tubes on ice.

5. Using a 12-channel pipette, add 33.3 μL *Hyb Cocktail* from the 12-strip PCR tubes to each well on the *Digest Plate*. Pipette up and down 10 times to mix.

6. Pour approximately 1 mL of H_2O into a 50 mL reagent reservoir.

7. Using a 12-channel pipette, add 65.7 μL H_2O from the reagent reservoir to each well on the *Digest Plate*. Pipette up and down 10 times to mix.

8. Seal the *Digest Plate* with clear adhesive film and spin the plate at 1,000 rpm for 15 s.

9. Place the *Digest Plate* on the thermal cycler and start the program *Meg Denature*.

10. Immediately following the completion of the *Meg Denature* program, place the *Digest Plate* on ice and cover with aluminum foil to prevent light exposure.

11. Let the *Digest Plate* cool for 2 min.

12. Spin the *Digest Plate* at 1,000 rpm for 15 s. Set the plate aside covered with aluminum foil.

**3.10. MIP Microarrays–
Preparing
the Microarrays
and Loading
the Microarrays
Samples (Post-amp)**

1. Set the GeneChip® Hybridization Oven 640 to 39°C. It is crucial that this type of oven is used due to the reagent's sensitivity to light. If an alternative oven is used, ensure it is enclosed and the inside of the oven is not exposed to light.

2. Remove the 96 arrays from the 4°C refrigerator. Remove the arrays from their packaging and label two arrays per sample. Keep in mind each sample has been labeled with dATP/dTTP and dCTP/dGTP meaning each sample is now located in two wells of the *Digest Plate*. It is ideal to label each array with the sample name and the two nucleotides it has been tagged with.

3. Allow the arrays to warm to room temperature with the face of the array facing down. Place the arrays on a soft surface to ensure the front window is not scratched.

4. While the arrays are warming to room temperature, insert a 200 μL pipette tip into the upper right septum of the array.

5. Set a single channel pipette to 90 μL.

6. Pipette 90 μL from each well of the *Digest Plate* and load the sample onto the appropriately labeled array through the lower left sample. It is helpful to hold the array while loading the sample and watch the front window fill with the sample. For more information regarding which sample is in each well of the *Digest Plate* refer to Steps 8–15 in **Section 3.2**.

7. Once all 96 samples have been loaded onto the appropriate array, remove the 200 μL pipette tip from the upper right septum.

8. Load the arrays onto the rotators in the GeneChip® Hybridization Oven which was previously set to 39°C. Set the rotator to 25 rpm.

9. Close the GeneChip® Hybridization Oven and leave the arrays in the oven for 12–16 h. The time should be kept consistent between experiments. Consistency is imperative.

**3.11. MIP Microarrays–
Staining, Washing,
and Scanning
the Microarrays
(Post-amp)**

1. Prepare the *Stain Cocktail* mixing the following reagents: 104.54 mL Wash A, 528 μL BSA (10 mg/mL), and 528 μL SAPE (1 mg/mL). This volume of *Stain Cocktail* is enough for 96 arrays plus an additional 10% material to account for pipette error.

2. For each array, 96 total, aliquot 360 μL *Stain Cocktail* to a 1.5 mL Eppendorf tube. This tube will be placed in position 1 on the GeneChip® Fluidics Station 450.

3. To prepare the GeneChip® Fluidics Station 450, load Wash A and Wash B onto the fluidics station. There is a position to hold each bottle of reagents on the right-hand side of the

machine. Each tube is labeled and should be submerged into the corresponding reagent. The waste bottle should be monitored and emptied when full.

4. Open the GCOS software and clock on the "Fluidics" button. Open the "Protocol" menu and select "Prime_450." The computer will prompt you to complete actions on the fluidics station. Follow the prompts.

5. When "Prime_450" has reached completion select the protocol in GCOS indicated by your Affymetrix representative to begin the staining and washing of the arrays.

6. Each GeneChip® Fluidics Station 450 will process only four arrays at a time. When prompted by the program load the first four arrays onto the fluidics station, the window of the array facing outwards. Do not push the lever down which will lock the array into place until prompted to do so by the program.

7. When prompted load the *Stain Cocktail* into position 1 on the GeneChip® Fluidics Station 450 and two empty 1.5 mL Eppendorf tubes into the other two positions. Again do not push the lever down which will bring the needles down into the Eppendorf tubes, wait until prompted to do so by the program.

8. Once the GeneChip® Fluidics Station 450 has loaded the first four arrays and lowered the needles into the Eppendorf tubes, the washing and staining of the arrays will begin.

9. The fluidics station should periodically be monitored to ensure there have been no errors. If errors occur it is best to contact Affymetrix technical support for help.

10. Once the first four arrays have been stained, washed, and the protocol reaches completion, load the next four arrays onto the fluidics station and again follow the prompts of the protocol. It is imperative a fresh Eppendorf tube of *Stain Cocktail* be loaded into position 1 for each array.

11. Staining and washing should be repeated for the first 48 arrays.

12. For the arrays that have been stained and washed, load each into the carousel of the Affymetrix GeneChip® Scanner. The arrays will only fit one way; do not force the arrays into the slots.

13. The arrays can be held in the scanner until the first 48 arrays have been stained and washed. The Affymetrix GeneChip® Scanner only holds 48 arrays.

14. When the first 48 arrays have been loaded onto the Affymetrix GeneChip® Scanner, begin the scanning by clicking on the "Start" button.

15. Repeat Steps 6–14 for the second batch of 48 arrays.

3.12. Data Analysis

1. GeneChip® Targeted Genotyping Analysis Software is used for analysis of the MIP data sets read from the arrays. Signal from each chip is background subtracted, color separated, normalized, and genotypes called as described previously within the software package (5).

2. Using normal diploid genome reference samples run at the same time as the tumor sample, the average signal in each of the three clusters (two homozygous clusters and a heterozygous cluster) for each marker as well as the standard deviation of the signal after removing (15%) outliers are calculated. The average signal in a cluster is then set to denote two copies since most reference normal samples will be diploid at any given point in the genome. For homozygous clusters we only consider the signal in the relevant allele and ignore the signal in the other allele for the computation of copy number. For heterozygous clusters we consider both signals and analyze them in two (orthogonal) directions: summing them together (copy sum analysis) and taking their ratio (allele ratio analysis). If a marker in a test sample has an allele imbalance, it may be classified as homozygous and therefore the signal in the other allele ignored.

4. Notes

1. It is imperative that the method remains consistent run to run. Consistency includes, but is not limited to, number of times pipetted, thermal cyclers used, pipetman, benches, etc. If the method and related procedures are not kept constant, the quality of data will be negatively affected.

2. Thermal cycling programs should be preprogrammed and properly named prior to begin experiments.

3. Two separated work areas should be created to minimize contamination.

4. Plates should be prelabeled prior to starting any run.

5. Never move any reagents, supplies, or equipment from the post-amp lab to the pre-amp lab; it is especially important that no PCR-amplified products are ever present in the pre-amp work area or brought into the pre-amp area. Each method will indicate the location the procedure should be completed.

Acknowledgments

We would like to thank Drs. Yuker Wang and Malek Faham for their contribution in developing MIP technology and providing information about the assay. This work was supported by NIH grants K08 CA96879, R21 CA109190, and 2P01HG000205.

References

1. Hardenbol, P., Yu, F., Belmont, J., Mackenzie, J., Bruckner, C., Brundage, T., Boudreau, A., Chow, S., Eberle, J., Erbilgin, A., Falkowski, M., Fitzgerald, R., Ghose, S., Iartchouk, O., Jain, M., Karlin-Neumann, G., Lu, X., Miao, X., Moore, B., Moorhead, M., Namsaraev, E., Pasternak, S., Prakash, E., Tran, K., Wang, Z., Jones, H. B., Davis, R. W., Willis, T. D., and Gibbs, R. A. (2005) Highly multiplexed molecular inversion probe genotyping: over 10,000 targeted SNPs genotyped in a single tube assay. *Genome Res* **15**, 269–275.

2. Ji, H., Kumm, J., Zhang, M., Farnam, K., Salari, K., Faham, M., Ford, J. M., and Davis, R. W. (2006) Molecular inversion probe analysis of gene copy alterations reveals distinct categories of colorectal carcinoma. *Cancer Res* **66**, 7910–7919.

3. Wang, Y., Moorhead, M., Karlin-Neumann, G., Falkowski, M., Chen, C., Siddiqui, F., Davis, R. W., Willis, T. D., and Faham, M. (2005) Allele quantification using molecular inversion probes (MIP). *Nucleic Acids Res* **33**, e183.

4. Wang, Y., Moorhead, M., Karlin-Neumann, G., Wang, N. J., Ireland, J., Lin, S., Chen, C., Heiser, L. M., Chin, K., Esserman, L., Gray, J. W., Spellman, P. T., and Faham, M. (2007) Analysis of molecular inversion probe performance for allele copy number determination. *Genome Biol* **8**, R246.

5. Moorhead, M., Hardenbol, P., Siddiqui, F., Falkowski, M., Bruckner, C., Ireland, J., Jones, H. B., Jain, M., Willis, T. D., and Faham, M. (2006) Optimal genotype determination in highly multiplexed SNP data. *Eur J Hum Genet* **14**, 207–215.

Chapter 7

A Whole-Genome Amplification Protocol for a Wide Variety of DNAs, Including Those from Formalin-Fixed and Paraffin-Embedded Tissue

Pamela L. Paris

Abstract

High-resolution genomic arrays and next-generation sequencers are some of the genome-based technologies poised to make significant contributions in the near future to basic and clinical science. The success of these technologies, and most certainly their translation into the clinic, will require that they produce high quality, reproducible data from small archived tumor specimens, including biopsies. DNA from patient samples, especially archival tissue, can be a limiting factor and lead to the need for amplification of the starting material. A variety of whole-genome amplification techniques are available, but choosing the most reliable, reproducible amplification technology that will be suitable for use across a wide spectrum of clinical specimens is essential. Sigma's whole-genome amplification kit provides a robust, highly reliable, and versatile amplification system across a variety of DNA sources. This chapter will detail Sigma's amplification protocol along with an optimized DNA extraction protocol for formalin-fixed and paraffin-embedded tissue.

Key words: Whole-genome amplification, clinical specimens, formalin-fixed paraffin-embedded, DNA extraction, array comparative genomic hybridization (aCGH).

1. Introduction

Identification of diagnostic and prognostic biomarkers using genome-based technologies is an important goal of research and a growing field where the scientist can choose from an array of technologies. Application of these discovery tools and their extension to the clinical setting will require high-quality data using limited DNA obtained from formalin-fixed and paraffin-embedded (FFPE) and small tumor specimens such as those from fine-needle aspirates and biopsies. However, the requirement for

Jonathan R. Pollack (ed.), *Microarray Analysis of the Physical Genome: Methods and Protocols, vol. 556*
© Humana Press, a part of Springer Science+Business Media, LLC 2009
DOI 10.1007/978-1-60327-192-9_7 Springerprotocols.com

microgram quantities of DNA for some of these whole-genome technologies (array comparative genomic hybridization (aCGH) copy number arrays *(1–3)*, single-nucleotide polymorphism (SNP) arrays *(4, 5)*, next-generation sequencing *(6, 7)*) can be a limiting factor. At present these applications demand a reliable, high-fidelity whole-genome amplification scheme to produce sufficient DNA.

Efforts have been made to design and optimize DNA amplification methodologies *(8–16)*. Various polymerase chain reaction (PCR)-based strategies, including degenerate oligonucleotide primed PCR (DOP-PCR) and linker-adapter (LA)-mediated PCR have demonstrated efficacy *(8–10)*. In DOP-PCR, a series of low temperature followed by higher-temperature annealing and extensions are performed using degenerate primers to amplify the genome globally and specifically. A disadvantage to DOP-PCR is that non-specific priming occurs and creates DNA with no relation to the original template, which becomes more pronounced when the amplified product is used in high-resolution applications. In LA-PCR, the template is enzymatically digested at a common restriction site, ligated to oligonucleotides, and amplified with primers specific to the linked oligonucleotides. This allows for little less sequence selection bias in LA-PCR than in DOP-PCR, but the procedure is more challenging. An improved thermal cycling protocol involving random non-enzymatic fragmentation and PCR amplification using ligated universal adapters has been developed and is commercially available (GenomePlex, now sold as Sigma's Whole Genome Amplification kit, WGA) *(13, 15)*. LA-PCR-based techniques, including Sigma's WGA, are well suited for degraded samples (e.g., FFPE) since the enzyme digestion occurs about once every 200–300 bp, and for high–molecular weight DNA (e.g., cell line) since the digestion step helps make the entire genome accessible.

Multiple displacement amplification (MDA) *(17)*, which utilizes the bacteriophage Phi29 DNA polymerase and random primers on long DNA templates (e.g., Qiagen's RepliG kit) or short, circularized DNA templates (rolling circle amplification (RCA) and restriction and circularization-aided (RCA), RCA–RCA), displaces new single-stranded DNA available for subsequent priming and isothermal amplification *(11, 14, 16)*. The amplification efficiency of MDA, which operates on long DNA templates, may be compromised for FFPE material which is often fragmented. Small DNA fragments in highly degraded FFPE samples could cross-ligate in RCA–RCA to form larger non-representative fragments that circularize and are amplified. The Phi29's extraordinary processivity reduces many of the sequence biases observed with PCR-based amplification, but also leads to the generation of extremely long products which can be difficult to work with downstream.

In our hands the most reliable, reproducible amplification technology that is suitable for use across a wide spectrum of research and clinical specimens (genomic DNA extracted from immortalized cell lines, blood, frozen and FFPE specimens) is Sigma's WGA kit *(18)*. Both sequence accuracy and gene dosage are important for use of an amplified product with aCGH, and therefore will be described here to demonstrate the fidelity of Sigma's amplification technology. The DNA extraction protocol is very important for FFPE specimens, so that will be detailed as well.

2. Materials

2.1. FFPE Sample Preparation

2.1.1. Macrodissection and Deparaffinization

1. #10 Feather scalpel (Fisher)
2. Small and large Kimwipes
3. 70% ethanol in a spray bottle
4. Zerostat-3 static gun (Sigma)
5. Hematoxylin and eosin (H&E) guide slide of cut section, with high-percentage tumor areas circled by a pathologist (*see* **Note 1**)
6. Hemo-De (Fisher) (*see* **Note 2**)

2.1.2. DNA Extraction

1. Gentra Puregene DNA Purification kit (Qiagen) (kit includes: Cell Lysis Solution, Protein Precipitation Solution, RNase A Solution, DNA Hydration Solution)
2. Proteinase K (PK), PCR grade ~20 µg/µl (Roche) (*see* **Note 3**)

2.1.3. DNA Clean-Up

1. Phase Lock Gel Light Tubes, 2 ml (Eppendorf)
2. Phenol:Chloroform:Isoamyl pH ~8 (Amresco)
3. Sodium acetate, 3 M, pH ~5.2, Molecular Biology Grade, (Sigma)
4. Agarose
5. 10X Tris-borate-EDTA (TBE) buffer
6. 1 kb ladder

2.2. DNA Amplification

1. GenomePlex Whole Genome Amplification kit (Sigma) (kit includes the following: 10X Fragmentation Buffer, 1X Library Preparation Buffer, Library Stabilization Solution, Library Preparation Enzyme, 10X Amplification Master Mix, nuclease-free water, control human genomic DNA).
2. JumpStart Taq DNA Polymerase (Sigma).

3. PCR tubes, individual with attached caps (*see* **Note 4**).

4. QIAquick PCR purification kit (Qiagen) (kit includes the following: Buffer PB, Buffer PE, Buffer EB, QIAquick spin columns, collection tubes).

3. Methods

An FFPE radical prostatectomy specimen was processed using the methods detailed in this chapter. The archived tissue block was 6 years old. The DNAs from the DNA extraction and amplification procedures are shown on a gel in **Fig. 7.1**. Our laboratory has shown that aCGH is a robust system that can provide consistent results from fragmented, relatively poor-quality DNA obtained from archived FFPE prostate tissue *(2, 3, 19, 20)*. The results of using the nonamplified and amplified DNA on aCGH are shown in **Fig. 7.2**. Any regional variation in amplification can be effectively normalized when test and reference genomes are both amplified and co-hybridized in an aCGH experiment *(18, 21)*.

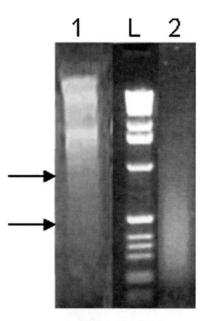

Fig. 7.1. Gel chromatography. DNA extracted from FFPE tissue according to described protocol (lane 1). Corresponding DNA after amplification with Sigma WGA kit (lane 2). *Arrows* denote 500 bp and 1,000 bp for the 1 kb ladder (L). Approximately 150 ng of each DNA was loaded on the 1.5% agarose gel, which was run for 1 h at 100 V.

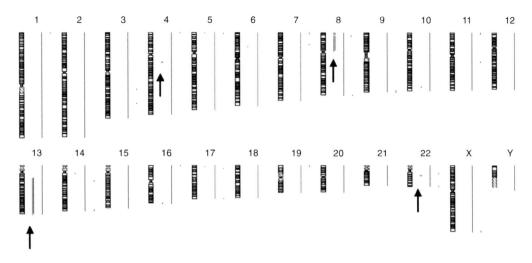

Fig. 7.2. aCGH results. Nonamplified (*light gray*) and amplified (*dark gray*) FFPE DNA were run on Agilent 244K oligonucleotide arrays. The ideogram for each chromosome is shown, along with copy number gains and losses to the *left* and *right*, respectively of the *vertical lines*. *Arrow heads* denote concordance between the amplified and nonamplified sample. No attempt has been made to filter the data for single copy number changes.

3.1. FFPE Sample Preparation

For this example, a radical prostatectomy case was obtained from the UCSF Tissue Core. Ten 10-µm slices were cut from the FFPE block and adhered to uncharged glass microscope slides. Tumor regions (>75%) on the H&E guide slides were outlined with the help of a pathologist.

3.1.1. Macrodissection and De-Paraffinization

1. The bench area should be cleaned with 70% ethanol. Gloves and a lab coat should be worn. Large Kimwipes are placed on the benchtop to create a workspace. The static gun is applied a few times to each unstained slide and an empty Eppendorf tube.

2. Macrodissection is performed with a scalpel, by placing the guide slide under the unstained slide and tracing the circled tumor area. The tissue removed from outlining is placed onto a small Kimwipe for disposal. Large strokes are used to scrape within the outlined area. The scalpel is used to compress the tissue scrapings into a group and then carefully transferred to a 1.5 ml Eppendorf tube. Sometimes one has to scrape the scalpel on the inside of the tube to dislodge the tissue. This is repeated for all of the unstained slides for this case (*see* **Note 5**).

3. Add 500 µl Hemo-De to the tube of tissue, vortex a bit on level 3–4, short spin (~10 s at 14,700g in a centrifuge), heat for 3–5 min in 55°C block, then place on a rotating Rocker and rock full speed for 5 min, then spin in a centrifuge for 10 min at 14,700g. Carefully, with a 200 µl pipette, remove the Hemo-De. For first two rinses, leave about 100 µl Hemo-De in the tube. Repeat this entire step twice. For the last rinse, remove all of the liquid (*see* **Note 6**).

4. Add 500 µl 100% ethanol, flick the tube, put on rocking platform for 5 min at full speed, then spin down in a centrifuge for 5 min at 14,700*g*. Remove most supernatant, and repeat. For the last rinse, remove all of the liquid (*see* **Note 6**). The resulting solid will be a little wet.

3.1.2. DNA Extraction

1. The Gentra (Minneapolis, MN) Puregene DNA Isolation kit is used to extract the DNA. Add 600 µl cell lysis solution from the **Gentra** kit. Vortex on level 2–4 briefly and then quick spin at 2,000*g* for 10 s. Add 10 µl Proteinase K (PK). Invert tube several times; do not vortex or enzyme will denature and lose activity. Parafilm, and place in a 55°C water bath. At end of day, vortex on level 2–4 briefly and then quick spin at 2,000*g* for 10 s. Add an additional 10 µl PK. Invert tube several times. Parafilm, and place in a 55°C water bath overnight.

2. Repeat PK for a total of six additions over 3 days. The solution should be fairly free of any particulates and therefore almost if not totally clear before proceeding. More PK can be added if it is not clear. Vortex on 2–4 briefly and then quick spin at 2,000*g* for 10 s.

3. Invert RNAse A supplied with Gentra kit to mix, and quick spin at 2,000*g* for 10 s. Add 3 µl RNAse A, invert 25 times, and quick spin at 2,000*g* for 10 s. Incubate at 37°C in water bath for 30 min. Cool sample to room temperature for 5 min.

4. Add 200 µl Protein Precipitation Solution (provided with Gentra kit) to the sample tube. Vortex at high speed for a full 20 s to mix uniformly. Put on ice for 5 min. Centrifuge at 14,700*g* for 10 min. A white pellet should be visible. Carefully remove most of supernatant (containing the DNA), without disturbing the protein pellet, and put it in a new 1.5 ml Eppendorf tube (*see* **Note 7**).

5. Add 600 µl 100% isopropanol (*see* **Note 8**). Mix by inverting gently 50 times. Centrifuge at 14,700*g* for 15 min, set at 20°C. Carefully remove most of the supernatant with a P-200 pipette, leaving 50–100 µl of the supernatant. Keep supernatants in case needed later (*see* **Note 9**).

6. Add 400 µl cold 70% ethanol. Invert several times. Spin 10 min at 14,700*g* at 20°C. Carefully remove most of the supernatant with a P-200 pipette. Spin 5 min at 14,700*g* at 20°C, and use a P20 to remove the last bit of supernatant. Let air dry for 15 min (*see* **Note 9**).

7. Add 100 µl of the DNA Rehydration Solution (provided with Gentra kit). Heat at 65°C for 15 min to help the DNA go into solution and inactivate any enzymes that might degrade the DNA. Put at 4°C overnight.

3.1.3. DNA Clean-Up

1. Let DNA solution warm to room temperature for 15–20 min.

2. "Pre-spin" an Eppendorf Phase Lock Gel Light tube at 14,700g for 1 min. Add the DNA solution (100 µl), and 100 µl room temperature phenol solution, in chemical hood wearing appropriate gloves, invert 10 times, then spin 5 min at 14,700g. Repeat (the addition of phenol, inverting, and spin). Put aqueous layer into a new, pre-spun Phase Lock Tube, repeat as above. Therefore, four phenol extractions will have been completed. Put final aqueous layer (∼ 100 µl) into new 1.5 ml tube.

3. Add 1/10 volume (10 µl) 3 M sodium acetate and 2X (220 µl) volume cold 100% ethanol. Invert tube 25 times or until DNA strands are visible. Put at −20°C overnight.

4. Centrifuge at 14,700g for 30 min at 4°C. Carefully remove most of the supernatant with a P-200 pipette, leaving 50–100 µl of the supernatant. Keep supernatants in case needed later (*see* **Note 9**).

5. Wash pellet with 200 µl cold 70% ethanol. Centrifuge at 14,700g for 15 min spin at 4°C. Carefully remove most of the supernatant with a P-200 pipette. Spin 5 min at 14,700g at 20°C, and use a P20 to remove the last bit of supernatant. Let air dry for 15 min (*see* **Note 9**).

6. Resuspend DNA in DNA Hydration solution (supplied with Gentra kit). If no pellet was ever visible, add 15–20 µl TE; for a small pellet, add 20–25 µl TE; for a DNA visible as strands during mixing, add 30–40 µl TE. Let sit overnight at 4°C.

7. Measure the DNA concentration with a UV-Vis spectrophotometer, such as the NanoDrop. In addition to a good yield, desire good quality which can be assessed by a 260/280 nm ratio of ∼1.8 and a 260/230 nm ratio of ∼2.0. The quality can further be assessed by running a 1.5% agarose gel.

3.2. DNA Amplification

A quantity of 100 ng of test and reference genomic DNA are individually amplified.

3.2.1. Fragmentation of Genomic DNA

1. Start with 100 ng DNA in a PCR tube; if volume is less than 10 µl, add water.

2. Add 1 µl 10X Fragmentation buffer (supplied with GenomePlex kit). Total reaction volume is 11 µl.

3. Heat at 95°C, for exactly 4 min. Immediately cool on ice. Centrifuge briefly to remove liquid from lid.

3.2.2. Creation of PCR-Amplifiable Library

1. Add 2 µl 1X Library Preparation Buffer and 1 µl Library Stabilization Solution (both supplied with GenomePlex kit). Pipet to mix. Centrifuge briefly.

2. Heat at 95°C for 2 min. Cool the sample on ice, briefly centrifuge, and put back on ice.

3. Add 1 μl Library Preparation Enzyme (supplied with GenomePlex kit). Pipet to mix, centrifuge briefly.

4. Run the following program on a thermocycler: 16°C for 20 min, 24°C for 20 min, 37°C for 20 min and 75°C for 5 min, followed by a 4°C hold.

5. Centrifuge briefly (*see* **Note 10**)

3.2.3. PCR Amplification

1. 7.5 μl 10X Amplification Master Mix (supplied with GenomePlex kit), 5 μl (12.5 units) Jumpstart Taq DNA polymerase and 47.5 μl nuclease free water is added to the sample. Pipet to mix, and centrifuge briefly.

2. Thermocycle as follows: Initial denaturation at 95°C for 3 min, 14 cycles of denaturing at 94°C for 15 s, and an annealing/extension step at 65°C for 5 min, and ending with a 4°C hold.

3.2.4. DNA Clean-Up

1. Add 25 μl water to bring DNA solution to 100 μl. Add 500 μl Buffer PB (supplied with QIAquick kit) and mix with pipette tip. Add the solution to a QIAquick column equipped with a collection tube.

2. Centrifuge for 1 min at 14,700*g*. Discard flow-through. Reuse collection tube.

3. Add 750 μl Buffer PE (supplied with QIAquick kit, to which ethanol has been added to the stock). Centrifuge for 30 s at 14,700*g*. Discard flow-through. Reuse collection tube. Centrifuge for 1 min at 16,000*g*.

4. Place column in a new 1.5 Eppendorf tube. Add 30 μl Buffer EB, incubate 5 min, and then centrifuge for 90 s at 14,700*g*. Let it sit overnight at 4°C.

5. The quality can be assessed by running 5 μl on a 1.5% agarose gel. The DNA should be in the 100–1,500 bp size range.

4. Notes

1. Assumes tissue is being cut and put onto slides in a tissue core, or the researcher has access to a microtome and the advice of a pathologist. If the tissue is fairly homogeneous, the tissue can be ribboned right into an Eppendorf and macrodissection skipped. For heterogeneous samples, macrodissect areas that are >75% tumor.

2. Hemo-De is a non-toxic xylene substitute that allows the work to be done on the bench top.

3. Although Proteinase K is supplied with the Gentra kit, the amount for complete digestion will exceed the volume provided. Roche's brand has always worked well for these digestions.

4. The use of individual PCR tubes with attached caps versus strip tubes is to avoid cross-contamination of samples with amplified products.

5. If other cases will be macrodissected, the Kimwipes, scalpel, and gloves should be discarded and replaced. The macrodissected tissue samples can be stored at room temperature until ready to process. If the DNA will be extracted the same day, the Hemo-De can be placed in the collection tube to make it easier to transfer the tissue from the scalpel to the tube.

6. For these last rinses of Hemo-De and ethanol, put the pipette tip flush to the bottom of the tube with the plunger depressed, then slowly release the plunger and you should be able to remove liquid and leave the solid behind. The solid does not adhere to the pipette tip.

7. Timing the vortexing and doing the spin with the centrifuge at 20°C helps guarantee success. If the protein pellet is diffuse, one can remove the supernatant to another tube, and respin.

8. Purchase smaller-sized isopropanol bottles in order to lessen the number of times the stock bottle is opened. Remove an aliquot only when needed and do not reuse.

9. DNA strands may or may not be visible, so note tube orientation in the centrifuge, and only pipet from the side opposite to where DNA should be. When removing supernatants from DNA, it is advisable to save the supernatant to a new tube in case DNA is inadvertently removed with the supernatant, which can be later determined and potentially recovered.

10. It is recommended to immediately go to the amplification step; however, Sigma claims the samples can be stored at −20°C up to 3 days.

References

1. Pinkel, D., Segraves, R., Sudar, D., Clark, S., Poole, I., Kowbel, D., Collins, C., Kuo, W.L., Chen, C., Zhai, Y. et al. (1998) High resolution analysis of DNA copy number variation using comparative genomic hybridization to microarrays. *Nat Genet*, **20**, 207–211.

2. Paris, P.L., Albertson, D.G., Alers, J.C., Andaya, A., Carroll, P., Fridlyand, J., Jain, A.N., Kamkar, S., Kowbel, D., Krijtenburg, P.J. et al. (2003) High-resolution analysis of paraffin-embedded and formalin-fixed prostate tumors using comparative genomic hybridization to genomic microarrays. *Am J Pathol*, **162**, 763–770.

3. Paris, P.L., Sridharan, S., Scheffer, A., Tsalenko, A., Bruhn, L. and Collins, C. (2007) High resolution oligonucleotide

CGH using DNA from archived prostate tissue. *Prostate*, **67**, 1447–1455.

4. Monzon, F.A., Hagenkord, J.M., Lyons-Weiler, M.A., Balani, J.P., Parwani, A.V., Sciulli, C.M., Li, J., Chandran, U.R., Bastacky, S.I. and Dhir, R. (2008) Whole genome SNP arrays as a potential diagnostic tool for the detection of characteristic chromosomal aberrations in renal epithelial tumors. *Mod Pathol*, **21**, 599–608.

5. Thompson, E.R., Herbert, S.C., Forrest, S.M. and Campbell, I.G. (2005) Whole genome SNP arrays using DNA derived from formalin-fixed, paraffin-embedded ovarian tumor tissue. *Hum Mutat*, **26**, 384–389.

6. Mardis, E.R. (2008) The impact of next-generation sequencing technology on genetics. *Trends Genet*, **24**, 133–141.

7. Schuster, S.C. (2008) Next-generation sequencing transforms today's biology. *Nat Methods*, **5**, 16–18.

8. Zhang, L., Cui, X., Schmitt, K., Hubert, R., Navidi, W. and Arnheim, N. (1992) Whole genome amplification from a single cell: implications for genetic analysis. *Proc Natl Acad Sci U S A*, **89**, 5847–5851.

9. Cheung, V.G. and Nelson, S.F. (1996) Whole genome amplification using a degenerate oligonucleotide primer allows hundreds of genotypes to be performed on less than one nanogram of genomic DNA. *Proc Natl Acad Sci U S A*, **93**, 14676–14679.

10. Klein, C.A., Schmidt-Kittler, O., Schardt, J.A., Pantel, K., Speicher, M.R. and Riethmuller, G. (1999) Comparative genomic hybridization, loss of heterozygosity, and DNA sequence analysis of single cells. *Proc Natl Acad Sci U S A*, **96**, 4494–4499.

11. Dean, F.B., Hosono, S., Fang, L., Wu, X., Faruqi, A.F., Bray-Ward, P., Sun, Z., Zong, Q., Du, Y., Du, J. et al. (2002) Comprehensive human genome amplification using multiple displacement amplification. *Proc Natl Acad Sci U S A*, **99**, 5261–5266.

12. Lage, J.M., Leamon, J.H., Pejovic, T., Hamann, S., Lacey, M., Dillon, D., Segraves, R., Vossbrinck, B., Gonzalez, A., Pinkel, D. et al. (2003) Whole genome analysis of genetic alterations in small DNA samples using hyperbranched strand displacement amplification and array-CGH. *Genome Res*, **13**, 294–307.

13. Little, S.E., Vuononvirta, R., Reis-Filho, J.S., Natrajan, R., Iravani, M., Fenwick, K., Mackay, A., Ashworth, A., Pritchard-Jones, K. and Jones, C. (2006) Array CGH using whole genome amplification of fresh-frozen and formalin-fixed, paraffin-embedded tumor DNA. *Genomics*, **87**, 298–306.

14. Lizardi, P.M., Huang, X., Zhu, Z., Bray-Ward, P., Thomas, D.C. and Ward, D.C. (1998) Mutation detection and single-molecule counting using isothermal rolling-circle amplification. *Nat Genet*, **19**, 225–232.

15. Pirker, C., Raidl, M., Steiner, E., Elbling, L., Holzmann, K., Spiegl-Kreinecker, S., Aubele, M., Grasl-Kraupp, B., Marosi, C., Micksche, M. et al. (2004) Whole genome amplification for CGH analysis: linker-adapter PCR as the method of choice for difficult and limited samples. *Cytometry A*, **61**, 26–34.

16. Wang, G., Maher, E., Brennan, C., Chin, L., Leo, C., Kaur, M., Zhu, P., Rook, M., Wolfe, J.L. and Makrigiorgos, G.M. (2004) DNA amplification method tolerant to sample degradation. *Genome Res*, **14**, 2357–2366.

17. Bagheri-Yarmand, R., Mazumdar, A., Sahin, A.A. and Kumar, R. (2006) LIM kinase 1 increases tumor metastasis of human breast cancer cells via regulation of the urokinase-type plasminogen activator system. *Int J Cancer*, **118**, 2703–2710.

18. Hittelman, A., Sridharan, S., Roy, R., Fridlyand, J., Loda, M., Collins, C. and Paris, P.L. (2007) Evaluation of whole genome amplification protocols for array and oligonucleotide CGH. *Diagnostic Molecular Pathology*, **16**, 198–206.

19. Paris, P.L., Andaya, A., Fridlyand, J., Jain, A.N., Weinberg, V., Kowbel, D., Brebner, J.H., Simko, J., Watson, J.E., Volik, S. et al. (2004) Whole genome scanning identifies genotypes associated with recurrence and metastasis in prostate tumors. *Hum Mol Genet*, **13**, 1303–1313.

20. van Dekken, H., Paris, P.L., Albertson, D.G., Alers, J.C., Andaya, A., Kowbel, D., van der Kwast, T.H., Pinkel, D., Schroder, F.H., Vissers, K.J. et al. (2004) Evaluation of genetic patterns in different tumor areas of intermediate-grade prostatic adenocarcinomas by high-resolution genomic array analysis. *Genes Chromosomes Cancer*, **39**, 249–256.

21. Huang, Q., Schantz, S.P., Rao, P.H., Mo, J., McCormick, S.A. and Chaganti, R.S. (2000) Improving degenerate oligonucleotide primed PCR-comparative genomic hybridization for analysis of DNA copy number changes in tumors. *Genes Chromosomes Cancer*, **28**, 395–403.

Chapter 8

Algorithms for Calling Gains and Losses in Array CGH Data

Pei Wang

Abstract

In this chapter, we introduce a few statistical algorithms for calling gains and losses in array-based comparative genomic hybridization (array CGH) data, including CBS, CLAC, CGHseg, and Fused Lasso. We illustrate the performance of the methods through simulated and real data examples. We also provide brief guidance on how to use the corresponding software at the end of this chapter.

Key words: CGH, calling gains, spatial structure, local correlation, false discovery rate (FDR).

1. Introduction

The output of array CGH experiments is usually a long vector, spanning each chromosome, recording the $\log_2$ ratios of normalized probe intensities from the test samples vs. the reference samples. These ratios of intensities are used to approximate the ratios of DNA copy numbers in the test samples vs. the reference samples. For those one-channel array platforms not using any reference samples, pseudo "reference" arrays with copy number exactly 2 are employed for calculating the $\log_2$ ratios. An example of such output is shown in **Fig. 8.1**. A positive (negative) $\log_2$ ratio indicates a possible DNA copy number gain (loss) for the probe. Our goal is to recover the true DNA copy numbers underlying variably noisy array CGH outputs and to identify the genome regions having copy number gains/losses. Since interpreting these results by eye is time consuming and not necessarily very accurate, in recent years, a number of algorithms have been developed to automatically call gains and losses from array CGH data.

Jonathan R. Pollack (ed.), *Microarray Analysis of the Physical Genome: Methods and Protocols, vol. 556*
© Humana Press, a part of Springer Science+Business Media, LLC 2009
DOI 10.1007/978-1-60327-192-9_8 Springerprotocols.com

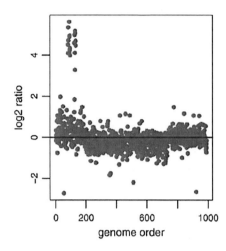

Fig. 8.1. A segment of CGH array for a glioblastoma multiforme (GBM) tumor. Each *grey dot* represents one gene/clone on the array. The *x*-axis represents the order of these gene/clones on the genome. The *y*-axis represents the $\log_2$ ratio of each gene/clone. The *grey line* is for $y = 0$ (i.e., normal copy number).

The few early methods addressing this problem were straightforward: smoothing the $\log_2$ ratio vectors followed by applying certain thresholds *(1–4)*. A common drawback of these methods is that the spatial relationship among genes on the genome – an important factor for DNA copy number alterations – is not taken into consideration. Here, the spatial correlation refers to the fact that two genes sitting close to each other on the genome are more likely to share the same DNA copy number than two genes sitting far away from each other. More specifically, it is often assumed that if we draw the true copy numbers along the genome, we shall get a stepwise curve.

To better capture the spatial structures, Fridlyand et al. proposed an unsupervised hidden Markov model (HMM) for identifying copy number changes on the chromosomes *(5)*. In this model, the underlying copy numbers are deemed as hidden discrete states with certain transition probabilities. However, the copy numbers measured in real applications always take continuous forms, as they are the average values of the copy numbers among millions of cells in the sample. Therefore, using only a few discrete states (usually three to five) to represent different levels of copy number alterations makes the HMM approach less efficient for complicated tumor samples.

Along the idea to divide the genomes into segments consisting of genes with the same copy numbers, Olshen and Venkatraman introduced another sophisticated method for array CGH analysis – "circular binary segmentation" (CBS) *(6)*. CBS employs the change-point model and modifies the original top-down binary segmentation strategy *(7)* to effectively detect change-intervals in the target genome regions. The change-point framework used in CBS provides a natural and elegant way for modeling array CGH data, which has been inherited by a few following works. For example, Zhang and Siegmund introduced a modified Bayesian information criterion (BIC) for the model selection in CBS *(8)*. The new criterion is derived from the model of Brownian motion with changing drift and has been demonstrated to work effectively on array CGH data. In addition, Lai et al. proposed a stochastic segmentation model which imposes Bayesian priors to both the changing locations and the changing levels *(9)*. The posterior distribution can then be used for confidence assessments of the segmentations.

In contrast to the top-down strategy employed by CBS, Wang etal. introduced a bottom-up agglomerative approach, "cluster along chromosomes" (CLAC) *(10)*, which enjoys better computational efficiency. CLAC builds hierarchical clustering-style trees along each chromosome arm (or chromosome), and then selects the "interesting" clusters (genome regions with copy number gains/losses) by controlling the false discovery rate (FDR) at a certain level.

A more "straightforward" segmentation approach is to directly search for the optimal break points on the genome to maximize some penalized likelihood function *(11–14)*. The choice of the penalty parameter/function in this process is extremely crucial. Picard et al. proposed to penalize the likelihood adaptively to the data and demonstrate the advantage of this approach over other model selection rules *(14)*.

While most segmentation methods employ parametric models for array CGH data, some non-parametric approaches that are free of distribution assumptions have also shown success in calling gains and losses in array CGH data. Li Hsu et al. *(15)* proposed to denoise the array CGH data using wavelet before making inferences on the aberrations. Tibshirani and Wang *(17)* developed a spatial smoothing approach using fused lasso regression *(16)* for calling gains and losses. The regression framework of fused lasso brings great computational efficiency and can be easily generalized to other analyses involving CGH data. The fused penalty (*see* **Section 2.4** for the definition) is also used to account for the spatial correlation in CGH data by Eilers and de Menezes *(18)* and Li and Ji *(19)*.

Other methods for calling gains and losses in array CGH data include a Bayes regression approach *(20)*, a pseudolikelihood approach *(21)* and others. A comprehensive comparison for some of these methods is given by Lai et al. *(22)*.

In the rest of this chapter, we focus on four methods: Olshen and Venkatraman (CBS) *(6)*, Wang et al. (CLAC) *(10)*, Picard et al. (CGHseg) *(14)*, and Tibshirani and Wang (cghFLasso) *(17)*, as these four methods are quite representative of the various approaches in this field and have all been implemented in sophisticated software that is publicly available. Specifically, in **Section 2**, we will explain the ideas and models of the four methods. In **Section 3**, we will illustrate the performance of the four methods with various data examples. In **Section 4**, we will give guidance on how to use the software.

2. Algorithms for Calling Gains and Losses

Denote the $\log_2$ ratios of the n genes (clones) on the targeted genome region as $\{x_1, x_2, \ldots, x_n\}$. Note that, we use the term "genes/clones" to refer to spots on CGH array chips, while there are some other studies using term "loci(locus)" instead, especially for BAC arrays. In addition, sometimes we may just use "gene" instead of "gene/clone" for simplicity.

2.1. Circular Binary Segmentation (CBS) (6)

Suppose the target genome region with n genes harbors one and only one amplified or deleted interval from the $i + 1^{\text{th}}$ gene to the jth gene. It is natural to assume that the measurements from the no-change region follow a normal distribution with mean 0 and standard deviation σ ($N(0, \sigma^2)$), while the measurements of the amplified/deleted interval follow a different normal distribution $N(\mu, \sigma^2)$. Here, the gene i and gene j are the change-points where the underlying copy number jumps. If i and j are given, the likelihood ratio test statistic for $H_0 : \mu = 0$ vs. $H_1 : \mu \neq 0$ (where H_0 and H_1 are the null and alternative hypothesis, respectively) can be easily calculated, and denoted as Z_{ij}. If the locations of the change-points are unknown, $Z_C = \max_{1 \leq i \leq j \leq n} |Z_{ij}|$ can be used for testing the null hypothesis of no change in the target genome region against the alternative that there is one change interval at an unknown location. The null hypothesis will be rejected when Z_C is large, and the corresponding change points will be estimated to be (i, j) satisfying $|Z_{ij}| = |Z_C|$. Here, if we paste the ends of this genome region together to form a circle, we can see that the circle consists of two arcs: the amplified/deleted arc and the normal arc. Thus, to test whether there is an alteration interval sitting on the target genome region is equivalent to testing whether the circle can be divided into two parts with different means. Then, following the idea of *binary segmentation procedure (7)*, CBS recursively applies the above test of Z_C to separate circles into two parts until no more changes are detected in any of the segments (circles) *(6)*.

A permutation approach is used in CBS for generating the null distribution of Z_C. After randomly permuting the order of the genes on the genome, the new CGH measurement vector shall contain no alteration interval, as the measurement of each gene is from the same distribution. Then, the corresponding $Z_C' = \max\left|Z_{ij}'\right|$ gives one realization of the test statistic Z_C under the null condition. An amplified/deleted interval will be declared if observed $|Z_C|$ exceeds the αth quantile of $|Z_C'|$ *(6)*.

Moreover, in order to avoid over-fitting, CBS performs a model selection procedure to choose the optimal number of segments after the recursive testing stops. Suppose there are K segments after the recursive testing. Denote the sum of squared deviations of each $\log_2$ ratio from their segment means as $SS(K)$. Consider all possible merging of two contiguous segments and calculate the sum of squared deviations of the resulting K–1 segments. Denote the smallest value as $SS(K-1)$. Repeat this process and calculate $SS(K-2), \cdots, SS(1)$. The optimal segment size is then defined as $k' = \min\{k : SS(k)/SS(K) < 1 + \gamma\}$, where γ is a tuning parameter (often set to be 0.05 or 0.10) *(6)*.

The computation cost of the above CBS algorithm is relatively high due to the large number of permutations conducted in the recursive testing process. To get around this difficulty, Venkatraman and Olshen *(23)* proposed to use the theoretical approximation of the tail probability of the test statistic Z_C derived by Siegmund *(24)* and Yao *(25)*. According to the asymptotical results *(24, 25)*, the theoretical approximations give good estimations of the true p-values when the number of genes in the target region is large. This leads to a hybrid approach which estimates p-values with permutation tests for middle and small regions, while estimates p-value with theoretical approximation for large regions. The authors suggested that this hybrid approach greatly improves the speed with minor loss in accuracy *(23)*.

An R version of the CBS algorithm has been implemented in the R package "*DNAcopy*".

2.2. Cluster Along Chromosome (CLAC) (10)

The CLAC algorithm uses a variation of the standard agglomerative clustering algorithm, a bottom-up strategy that generates a binary tree to represent the similarities in the data. The standard agglomerative clustering algorithm begins with every observation representing a singleton cluster and then recursively merging the closest two clusters until one root is left. Building a clustering tree along one chromosome arm (or one chromosome) differs in two ways from the standard agglomerative clustering. First, the order of the genes on the chromosome is fixed, i.e., the order of the leaves of the tree is fixed. So, only adjacent clusters are joined together when the tree is generated. Second, the "similarity" between two clusters no longer refers to the spatial distance but to the similarity of the array measurements ($\log_2$ ratio) between the two clusters. Denote the similarity measurement between the $\log_2$ ratios of two genes or two contiguous clusters as *rd* (relative difference). The clustering procedure on one chromosome arm can be summarized as follows: (1) begin with n clusters with one gene (clone) in each cluster; (2) merge the two adjacent clusters with the smallest rd; and (3) repeat Step 2 until one big cluster is obtained. The tree structure for a simulated example is shown in **Fig. 8.2**. The height of a node in the tree represents the rd between the left branch and right branch of that node. We can see that the amplification and deletion regions on the simulated chromosome can be represented by different branches in the tree, and the regions with copy number gains/losses have joined the tree at rather smaller rd levels than the noisy parts.

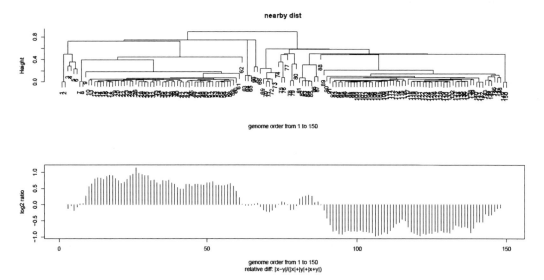

Fig. 8.2. CLAC tree structure for a simulated example. The *lower panel* shows the $\log_2$ ratio of CGH measurements for a chromosome arm of length 150, which is simulated as follows: (1) *Left*, amplified region: $g_i \sim N(0.7, 0.3)$, $j \in \{10, 11, \ldots 60\}$; (2) *Right*, deleted region: $g_i \sim N(-0.7, 0.3)$, $j \in \{90, 91, \ldots, 140\}$; (3) *Center*, noise region: $g_i \sim N(0, 0.3)$, $j \notin A \cup D$. An average smoothing of window-size 5 has been performed on this chromosome segment before the CLAC tree (*upper panel*) is built on it.

After we build the hierarchical tree, we need to decide which "clusters" are "interesting". We consider all $n-1$ clusters corresponding to all the branches in the tree consisting of at least two genes/clones. We examine three properties of each node/cluster: (1) *rd*: the height of this node in the tree (the biggest rd for nearby gene pairs in the cluster); (2) *size*: the size of the subtree with this node as the root (the number of genes in the cluster); and (3) *meanvalue*: the mean value of the leaves of the subtree (the mean value of the $\log_2$ ratio for genes in the cluster). Now we can use (*size, rd, meanvalue*) to select "interesting" regions. There are two different kinds of interesting regions. The first kind is characterized as a big spike, which is always a small region with extremely large positive or negative log ratio values (corresponding to high-level amplifications and homozygous deletions, respectively). The second kind is the consistent gain/loss region, whose log ratios might not deviate away from 0 very much, but tend to stay positive(gain) or negative(loss) in the whole region. The big spike regions correspond to the nodes with big |*meanvalue*|. The consistent gain/loss regions correspond to the nodes with bigger *size*, smaller *rd*, and with |*meanvalue*| not too small. Thus, candidate nodes representing alteration regions can be selected based on the joint distribution of (*size, rd, meanvalue*). The detailed selection rules are described in Wang et al. *(10)*.

In the end, a false discovery rate (FDR) is estimated for assessing the confidence of the called results. FDR is the expected proportion of rejected null hypotheses that are actually true *(26)*. Here, the null hypothesis for gene/clone i is that gene/clone i does not belong to any gain/loss region. The FDR can then be estimated using reference arrays as follows *(27–30)*

$$\text{FDR} = \frac{\text{number of genes picked in the reference array (under the same criteria)}}{\text{number of genes picked in the tumor array}}.$$

By varying the selection rules based on (*size, rd, meanvalue*), we are able to control the FDR under a target level.

CLAC method is conceptually simple and automatic, allowing estimation of its FDR over an entire array. It has been implemented in an R-package *clac* as well as an Excel add-in package **CGH-Miner**.

2.3. Penalized Likelihood Approach with Adaptive Model Selection Criteria (CGHseg) (14)

Suppose the target genome region consists of K segments with the true DNA copy numbers in each segment being the same. It is reasonable to assume the $\log_2$ measurements for the genes from the same segment are independently identically distributed normal. Then, the joint log-likelihood for all the data of the target genome region can be easily written out, and the segmentation that maximizes this joint likelihood gives a reasonable estimation of the true break points. If the number of segments K is given, dynamic programming can be used to seek the optimal break point locations maximizing the joint log-likelihood (*see* Picard et al. *(14)* for more details). However, the optimal number of segments (K) cannot be selected using the likelihood itself, for the likelihood increases with the model complexity (the number of segments). To address this problem, Picard et al. *(14)* proposed an adaptive approach for selecting the best K. For simplicity, denote the maximum value of the joint log-likelihood with k segments as $\widehat{L_k}$. The idea is to select k such that (1) the increase from $\widehat{L_{k-1}}$ to $\widehat{L_k}$ is significant, and (2) the increase from $\widehat{L_{m-1}}$ to $\widehat{L_m}$ is not significant for any m bigger than k. If we plot $\widehat{L_k}$ versus k, which generally gives a concave curve, then the inflexion point of this curve more or less possesses properties (1) and (2). Thus, Picard et al. *(14)* proposed to use $\hat{k} = \max\{k \in [1, n]| D_k < - n/2\}$, where $D_k = L_{k-1} - 2L_k + L_{k+1}$ approximates the second derivative of $\widehat{L_k}$ along k. As this procedure is adaptive to the data, it outperforms other model selection criteria using fixed parameters.

The method is implemented in MATLAB Software ***CGHseg***.

2.4. Spatial Smooth with Fused Lasso (cghFLasso) (15)

Tibshirani and Wang *(15)* proposed to use the "fused lasso" regression *(16)* for the detection of regions of gains or losses in CGH data. The fused lasso is a regularized regression technique, which seeks coefficients minimizing a loss function consisting of three terms: the sum of square error, the sum of absolute value of regression coefficients (the lasso penalty) *(31)*, and the sum of the absolute *difference* between contiguous coefficients (the fused penalty) *(16)*. The fused lasso is developed for situations when predictors (X) in the regression model have some kind of natural ordering. The lasso penalty controls the total number of nonzero coefficients in the model *(31)*. The fused penalty encourages the flatness of the coefficient profiles β_j as a function of j *(16)*. Here we apply fused lasso to recover the true copy number signal through smoothing the sequence $X = \{x_1, x_2, \ldots x_i, \ldots x_n\}$ along the one-dimensional index i. Since the CGH signal is approximated by a piecewise function that has relatively sparse areas with nonzero values, the fused lasso penalty is useful for determining which areas of the signal are likely to be nonzero. This is a convex optimization problem that can be efficiently solved with the *pathwise coordinate algorithm (32)*. After the underlying copy numbers are estimated with fused lasso regression, they will be further thresholded by a value θ to obtain the final regions of gains or losses. The choice of θ is used to control the FDR of the result. Again the FDR can be estimated using the reference arrays as described in the earlier equation (*see* **Section 2.2**). More details of the method are available in Tibshirani and Wang *(15)*.

Fused lasso can be deemed as a "smoothing" approach, for the fused penalty term in the loss function can be viewed as the first derivative of the coefficient profiles. However, we want to point out that the general smoothing methods are not typically useful for analyzing CGH data because their results can be difficult to interpret. This is illustrated in **Fig. 8.4**, where two popular smoothing methods – *lowess (33)* and *penalized smoothing splines (34)* – are applied to the example data (*see* **Section 3.2** for details). We used R function *lowess* (smooth window = 10) and *spm* (default parameters) to compute the results. As we can see from the figure, the two smoothing methods do not provide direct calls for copy number gains/losses and thus require additional thresholding for identifying regions with significant alterations. Moreover, the smoothing curves do not catch the piecewise constant shape of copy number changes, which raises additional challenges for controlling the FDR. Furthermore, copy number alterations can be both large chromosome region gains/losses and also abrupt local amplifications/deletions. Therefore, different degrees of smoothness are needed for different genome segments, which adds another layer of complexity to the kernel- and spline-based approaches. However, in fused lasso, the use of L1-norm on the fused penalty term enables the method to capture both the piecewise flatness patterns and the abrupt local jumps at the same time (*see* **Fig. 8.4**). In addition, the control on the overall sparsity of the coefficient solution helps to screen away the "cold"-spot regions (i.e., regions with no alterations). These make the fused lasso a more attractive approach for analyzing CGH data.

An R version of this method has been implemented in R package "*cghFLasso*".

3. Data Examples

In this section, we illustrate the performance of the four methods CBS (R package **DNAcopy**) *(6, 23)*, CLAC (R package *clac*) *(10)*, Penalized likelihood (Matlab package **CGHseg**) *(14)*, and Fused Lasso (R package *cghFLasso*) *(17)* on both simulated examples and arrays from real applications.

3.1. Simulation Studies

We first investigate the methods using a set of artificial CGH data simulated by Lai et al. *(22)* (available at http://www.chip.org/~ppark/supplement.html). We consider the most challenging situation where the signal-to-noise ratio is equal to 1 (equivalent to very noisy data). Each of the artificial chromosomes contains 100 probes, with one amplified region embedded in the middle. The width of the amplified region has four different possible sizes: 5, 10, 20, and 40. The smaller the amplified region is, the more difficult it is to correctly detect the copy number changes. There are 100 independent chromosomes simulated for each width size. We estimate the copy number with the various methods for all the chromosomes. The receiver-operator characteristic (ROC) curves of different methods are shown in **Fig. 8.3**, which suggest that fused lasso better captures the true DNA copy number alterations than the other three methods, especially when the aberration width is small.

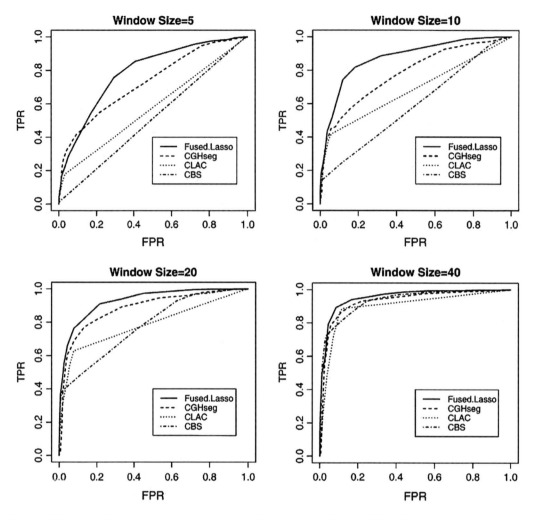

Fig. 8.3. ROC curves of the four methods on the simulated dataset (*see* **Section 3.1**). TPR = (the number of probes within the aberration width that is above a threshold)/(the total number of probes within the aberration width); FPR = (the number of probes outside the aberration width that is above a threshold)/(the total number of probes outside the aberration width).

3.2. Real Applications

(a) *GBM data:* The glioblastoma multiforme (GBM) data from Bredel et al. *(35)* contain samples representing primary GBMs, a particular malignant type of brain tumor. We investigate the performance of various methods on the array CGH profiles of these GBM samples. To generate a more challenging situation where both local amplification and large region loss exist on the same chromosome, we paste together the following two array regions: (1) chromosome 7 in GBM29 from 40 to 65 Mb and (2) chromosome 13 in GBM31. The performance of different methods on this pseudo chromosome is illustrated in **Fig. 8.4**. We can see that all four methods successfully identified both the local amplification and the big chunk of copy number loss.

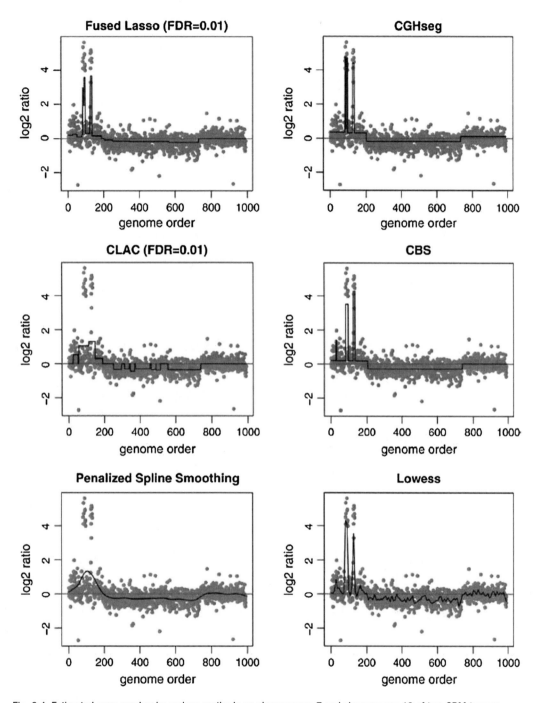

Fig. 8.4. Estimated copy number by various methods on chromosome 7 and chromosome 13 of two GBM tumors.

(b) *Breast tumor data:* In the study conducted by Pollack et al. *(1)*, cDNA microarray CGH was used to profile across 6,691 mapped human genes in 44 breast tumor samples and 10 breast cancer cell lines. The scanned raw data was available from the Stanford Microarray Database (http://smd.stanford. edu). We pick the breast cancer cell line (MDA157) as an example, which has a large degree of copy number alterations. We apply the four methods on this cell line to estimate the underlying copy number changes. From the results shown in **Fig. 8.5**, we can see that (1) Fused Lasso successfully recognizes various copy number alterations; (2) CGHseg appears to be overly sensitive to outlier measurements and thus will be more suitable for detecting single-gene copy number changes

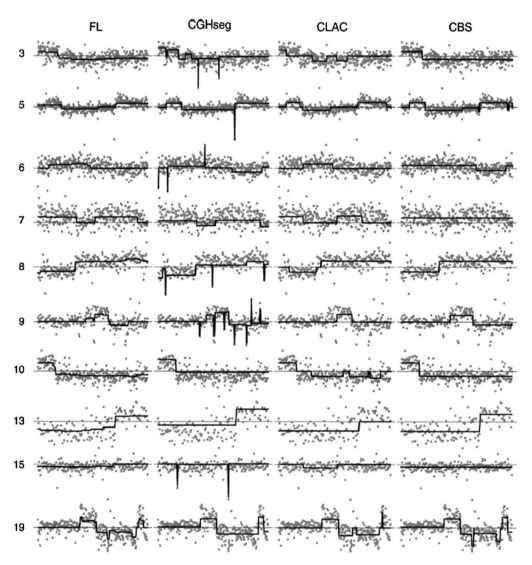

Fig. 8.5. Array CGH profiles of 10 chromosomes of breast cancer cell line MDA157. Panels in the same row are for the same chromosome. The integers at the beginning of each row are chromosome indexes. The *solid line* in each panel represents the estimated copy number of a particular method, whose name is shown on the *top* of each column. The *grey horizontal line* in each panel represents $y = 0$.

from high-quality arrays; (3) CLAC is conservative in handling outliers with opposite signs in the same alteration region and therefore tends to break large alteration segments into small blocks; (4) CBS provides clean solutions for segmentations but has the limitation to detect break points whose alteration signals are weak (e.g., chromosome 7 and 15 of the selected cell line).

3.3. Speed Comparison

We further investigate the computational efficiency of the four softwares on the breast cancer array CGH data *(1)*. For a given length n ($= 100$, 500, $1,000$, $2,000$, or $5,000$), we first randomly sample 50 genome segments of length n from the breast cancer CGH arrays. We then process the 50 segments with various softwares and summarize the CPU time in **Table 8.1**. From the numbers, we can see that *clac* and *cghFLasso* are computationally more efficient and thus are more suitable for processing high-density arrays with hundreds of thousands of genes/probes.

Table 8.1
Mean and the standard deviation (in the parenthesis) of the CPU time (seconds; on a server with two Dual/Core, CPU 3 GHz and 6 GB RAM) for various methods on simulated data

	$n=100$	$n=500$	$n=1,000$	$n=2,000$	$n=5,000$
DNAcopy	0.151 (0.113)	1.243 (0.804)	3.669 (1.135)	8.455 (2.854)	14.023 (8.422)
CGHseg	0.063 (0.008)	0.445 (0.016)	1.223 (0.041)	4.205 (0.104)	37.94 (0.621)
clac	0.049 (0.003)	0.086 (0.013)	0.157 (0.037)	0.368 (0.073)	0.965 (0.296)
cghFLasso	0.025 (0.013)	0.140 (0.017)	0.334 (0.036)	0.840 (0.056)	2.5814 (0.331)

Note: We have used the default parameters provided in the software on the above data examples. Users who are familiar with these methods may choose different parameters to achieve better performance.

4. Software

In this section, we describe the usage of *DNAcopy* (*CBS*), *clac*, *CGH-Miner*, *CGHseg*, and *cghFLasso*. To use these packages, you need to have **R** or **Matlab** (for *CGHseg* only) installed on your system. **R** is a free programming software that can be downloaded from http://cran.r-project.org/. **Matlab** can be purchased at http://www.mathworks.com/products/matlab/tryit.html. Both websites provide tutorial documents for basic **R/Matlab** programming.

4.1. DNAcopy

The homepage of *DNAcopy* is: http://bioconductor.org/packages/2.1/bioc/html/DNAcopy.html.

- *Installation:* After starting **R**, type the following code in the *RGui* command window:

  > Source("http://bioconductor.org/biocLite.R")

  > BiocLite("DNAcopy")

- *Usage of DNAcopy:* Below is an example code

```
> library(DNAcopy)
> CNA.object=CNA(arraydata, chrom, maploc, data.type=
    "logratio")
> smoothed.CNA.object=smooth.CNA(CNA.object)
> segment.CNA.object= segment ( smoothed.CNA.object,
    verbose=1)
> plot (segment.CNA.object, plot.type="w")
```

where "arraydata" is a vector representing the CGH measurements ($\log_2$ ratios) on the target genome; "chrom" and "maploc" are vectors showing the chromosome numbers and genome locations of the genes/clones in arraydata. Function "segment" is the main function performing the CBS analysis. The output is a data frame with each row representing one "segment" of the target genome, and six columns representing: (1) the sample id, (2) the chromosome number of the segment, (3) the map location of the start of the segment, (4) the map location of the end of the segment, (5) the number of markers in the segment, and (6) the average value in the segment. Type "help(segment)" to see more details.

4.2. clac and CGH-Miner

The homepage of *clac* and *CGH-Miner* is http://www-stat.stanford.edu/~wp57/CGH-Miner/.

- *Installation of clac:* After starting **R**, type the following code in the *RGui* command window:

  > install.packages("clac",dependencies=TRUE)

- *Usage of clac:* In below is an example code

```
> library(clac)
> NormalResult<-clac.preparenormal.R(DiseaseArray, Nor-
    mal Array, Normal.Type, chromosome.number, nucleo-
    tide. position)
> clac.result<-clac.tumorarray.R(NormalResult, tumorar-
    ray Index=1:n)
> clac.PlotSingleArray.R(1, NormalResult, clac.result)
> clac.PlotConsensus.R(clac.result, chromosome, nucposi-
    tion, 1:n)
```

where "DiseaseArray" is the data matrix recording the $\log_2$ ratios of disease samples with each column representing one array (sample) and each row representing one gene/clone; "NormalArray" is a similar data matrix recoding the $\log_2$ ratios of reference samples; "Normal.Type" is a vector indicating whether each reference array is from a same-gender hybridization or from a male/female hybridization (for the latter, X and Y chromosome data are excluded) ; "chromosome.number" and "nucleotide.position" are two vectors representing the chromosome number and the nucleotide position for each gene/clone on the array. Function "clac.tumorarray.R" is the main function performing CLAC analysis on all selected disease arrays (specified by "tumorarrayIndex"). The returned value is a list with two components: "Region Mean" and "fdr". "RegionMean" is a data matrix having the same dimension as the original input data matrix "DiseaseArray" of function "clac.preparenormal.R". For each entry of the matrix, 0 is reported if the gene/clone is not being called to have any copy number alteration, otherwise the amplification/deletion region mean of the genes/clones is reported. "fdr" is a numeric vector reporting the result FDR for each target disease array. See "help(clac.preparenormal.R)" or "help (clac.tumorarray.R)" for more details.

Installation and Usage of CGH-Miner. *CGH-Miner* is an Excel front-end of the CLAC method. It can be freely downloaded from http://www-stat.stanford.edu/~wp57/CGH-Miner/. For installation, click on the *CGH-Miner.exe* file and follow the instructions. Then *CGH-Miner* will be automatically available the next time you start up Excel. To run *CGH-Miner,* first format the data in an Excel sheet as follows: (1) the first row gives the column labels (Gene/Clone ID, Gene/Clone Name, ChromosomeNumber, NucleotidePosition, CGH-array); (2) the second row gives the class labels (NN – reference array from same-gender hybridization, NMF – reference array from male–female hybridization, Disease – target array to run analysis); (3) the third row indicates the sample labels for user's reference; and (4) the remaining lines contain CGH array measurements one line per gene/clone. Then click on CGH-Miner button in the menu bar and a dialog form will pop up. The user needs to specify several input parameters including the row numbers for Class Labels, Sample Labels, Data Starts, and the array type of the dataset. Click the OK button and the CLAC analysis will be performed. After the analysis is done, *CGH-Miner* adds three more worksheets to the workbook, recording the analysis results as well as summary figures. More details are described in the manual (http://www-stat.stanford.edu/~wp57/CGH-Miner/CGH-Miner-Manual.pdf).

4.3. CGHseg

The Matlab package *CGHseg* can be downloaded from http://www.agroparistech.fr/mia/outil.html.

- *Usage of CGHseg*: After starting **Matlab**, type "CGH_segmentation" in the command window. Then the user will be asked to specify the path and the name of the input file. The table below gives an example, where the words in bold are typed by the user:

> **CGH_segmentation**
Path directory : (default :"C:\MATLABR11\work\") **C:\CGH_segmentation**
Name of the input data file (.txt): **Input.txt**
Specifying options? ('N' : model homoscedastic, selection Marc Lavielle, Km=20, graphes, result.txt) (N/Y) : **N**
Which chromosome ?(write the n?or 'all') **all**

The input file gives the data matrix of one CGH array: each row is for one gene/clone with four numbers representing the gene/clone name, $\log_2$ ratio, nucleotide position in kb, and chromosome number of this gene/clone (no column name in the file), respectively. The output of the program will be saved in a file called "result.txt" in the same directory as the input file. This file contains a data matrix with each row representing one inferred segment, and seven columns representing (1) chromosome of the segment; (2) beginning breakpoint coordinate of the segment; (3) ending breakpoint coordinate; (4) beginning gene/clone name; (5) ending gene/clone name; (6) mean $\log_2$ ratio for the segment; and (7) standard deviation of the $\log_2$ ratio of the segment.

4.4. cghFLasso

The homepage of *cghFLasso* is: http://www-stat.stanford.edu/~tibs/cghFLasso.html.

- *Installation*: After starting **R**, type the following code in the *RGui* command window:
  > install.packages("cghFLasso",dependencies = TRUE)
- *Usage of cghFLasso*: In below is an example code

> library(cghFLasso)
> Normal.FL<-cghFLasso.ref(Normal.Array, chromosome)
> Disease.FL<-cghFLasso(Diease.Array, chromosome, nucleotide.position, FL.norm=Normal.FL, FDR=0.1)
> plot(Disease.FL, index=i, type="Single")
> plot(Disease.FL, index=1:4, type="Consensus")

where "Disease.Array" is the data matrix recording the $\log_2$ ratios of disease samples with each column representing one array (sample) and each row representing one gene/clone; "Normal.Array" is a similar data matrix recoding the $\log_2$ ratios of reference samples; "chromosome" and "nucleotide.position" are two vectors representing the chromosome number and the nucleotide position for each gene/clone on the array; "FDR" is the user-specified False discovery level of the analysis. Function "cghFLasso" is the main function estimating the underlying copy numbers on all selected disease arrays. The returned list has one component (Esti. CopyN) reporting the estimated DNA copy numbers. See "help (cghFLasso)" for more details.

References

1. Pollack, J., Sorlie, T., Perou, C., Rees, C., Jeffrey, S., Lonning, P., Tibshirani, R., Botstein, D., Borresen-dale, A. and Brown, P. (2002). Microarray analysis reveals a major direct role of DNA copy number alteration in the transcriptional program of human breast tumors. *Proc. Natl. Acad. Sci. USA* **99**, 12963–12968.

2. Hodgson, G., Hager, J., Volik, S., Hariono, S., Wernick, M., Moore, D., Nowak, N., Albertson, D., Pinkel, D., Collins, C., Hanahan, D. and Gray, J.W. (2001). Genome scanning with array CGH delineates regional alterations in mouse islet carcinomas. *Nat. Genet.* **29**, 491.

3. Cheng, C., Kimmel, R., Nelman, P. and Zhao, L.P. (2003). Array rank order regression analysis for the detection of gene copy-number changes in human cancer. *Genomics* **82**, 122–129.

4. Lingjaerde, O., Baumbusch, L., Liestol, K., Glad, I. and AL, B.-D. (2005). CGH-explorer: a program for analysis of array-CGH data. *Bioinformatics* **21**(6).

5. Fridlyand, J., Snijders, A.M., Pinkel, D., Albertson, D.G. and Jain, A.N. (2004). Hidden Markov models approach to the analysis of array CGH data. *J. Multivariate Anal.* **90**, 132–153.

6. Olshen, A. and Venkatraman, E. (2004). Circular binary segmentation for the analysis of array-based DNA copy number data. *Biostatistics* **5**, 557–572.

7. Vostrikova, L.J. (1981). Detecting 'disorder' in multidimensional random processes. *Sov. Math. Dokl.* **24**, 55–59.

8. Zhang, N.R. and Siegmund, D.O. (2007). A modified Bayes information criterion with applications to the analysis of comparative genomic hybridization data. *Biometrics* **63**, 22–32.

9. Lai, T.L., Xing, H.P. and Zhang, N.R. (2007). Stochastic segmentation models for array-based comparative genomic hybridization data analysis. *Biostatistics*, doi:10.1093/biostatistics/kxm031

10. Wang, P., Kim, Y., Pollack, J., Narasimhan, B. and Tibshirani, R. (2005). A method for calling gains and losses in array CGH data. *Biostatistics* **6**, 45–58.

11. Myers, C.L., Dunham, M.J., Kung, S.Y. and Troyanskaya, O.G. (2004). Accurate detection of aneuploidies in array CGH and gene expression microarray data. *Bioinformatics* **20**, 3533–3543.

12. Lipson, D., Aumann, Y., Ben-Dor, A., Linial, N. and Yakhini, Z. (2005). Efficient calculation of interval scores for DNA copy number data analysis. In *Proceedings of RECOMB 05*. Springer-Verlag, Cambridge, MA.

13. Hupe, P., Stransky, N., Thiery, J.-P., Radvanyi, F. and Barillot, E. (2004). Analysis of array CGH data: from signal ratio to gain and loss of DNA regions. *Bioinformatics*. **20**, 3413–3422.

14. Picard, F., Robin, S., Lavielle, M., Vaisse, C. and Daudin, J.J. (2005). A statistical approach for array CGH data analysis. *BMC Bioinform.* **11**, 6–27.

15. Hsu, L., Self, S.G., Grove, D., Randolph, T., Wang, K., Delrow, J.J., Loo, L. and Porter, P. (2005). Denoising array-based comparative genomic hybridization data using wavelets. *Biostatistics*. **6**, 211–226.

16. Tibshirani, R. and Wang, P. (2007). Spatial smoothing and hot spot detection for CGH data using the fused lasso. *Biostatistics*, doi:10.1093/biostatistics/kxm013.

17. Tibshirani, R., Saunders, M., Rosset, S., Zhu, J. and Knight, K. (2004). Sparsity and smoothness via the fused lasso. *J. R. Stat. Soc. B.* **67**(1), 91–108.

18. Eilers, P.H. and de Menezes, R.X. (2005). Quantile smoothing of array CGH data. *Bioinformatics* **21**(7), 1146–1153.

19. Li, Y. and Zhu, J. (2007). Analysis of array CGH data for cancer studies using the fused quantile regression. *Bioinformatics* **23**, 2470–2476.

20. Wen, C., Wu, Y., Huang, Y., Chen, W., Liu, S., Jiang, S., Juang, J., Lin, C., Fang, W., Hsiung, C. and Chang, I. (2006). A Bayes regression approach to array-CGH data. *Stat. Appl. Mol. Biol. Berkeley Electron. Press* **5**(1), 3.

21. Engler, D., Mohapatra, G., Louis, D. and Betensky, R. (2006). A pseudolikelihood approach for simultaneous analysis of array comparative genomic hybridizations. *Biostatistics* 7(3), 399–421.

22. Lai, W.R., Johnson, M.D., Kucherlapati, R. and Park, P.J.(2005). Comparative analysis of algorithms for identifying amplifications and deletions in array CGH data. *Bioinformatics* **21**(19), 3763–3770.

23. Venkatraman, E.S. and Olshen, A.B. (2007). A faster circular binary segmentation algorithm for the analysis of array CGH data. *Bioinformatics* **23** (6), 657–663.

24. Siegmund, D.O. (1988). Approximate tail probabilities for the maxima of some random fields. *Ann. Probab.* **16**, 487–501.

25. Yao, Q. (1989). Large deviations for boundary crossing probabilities of some random fields. *J. Math. Res. Exposit.* **9**, 181–192.

26. Hastie, T., Tibshirani, R. and Friedman, J. (2001). *The Elements of Statistical Learning.* Springer, New York, NY, p. 475.

27. Benjamini, Y. and Hochberg, Y. (1995). Controlling the false discovery rate: a practical and powerful approach to multiple testing. *J. R. Stat. Soc. B* **57**(1), 289–300.

28. Tusher, V., Tibshirani, R. and Chu, G. (2001). Significance analysis of microarrays applied to the ionizing radiation response. *Proc. Natl. Acad. Sci. USA* **98**, 5116–5121.

29. Storey, J. (2002). A direct approach to false discovery rates. *J. R. Stat. Soc.* **64**(3), 479–498.

30. Efron, B. and Tibshirani, R. (2002). Microarrays, empirical Bayes methods, and false discovery rates. *Genetic Epidemiology* **23**(1), 70–86.

31. Tibshirani, B. (1996). Regression shrinkage and selection via the lasso. *J. R. Stat. Soc. Ser. B* **58**, 267–288.

32. Friedman, J., Hastie, T. and Tibshirani, R. (2007). Pathwise coordinate optimization. *Ann. Appl. Stat.* **1**(2), 302–332.

33. Becker, R.A., Chambers, J.M. and Wilks, A.R. (1988). *The New S Language.* Wadsworth Brooks Cole, Pacific Grove, CA.

34. Ruppert, D., Wand, M.P. and Carroll, R. (2003). *Semiparametric Regression.* Cambridge University Press, New York.

35. Bredel, M., Bredel, C., Juric, D., Harsh, G.R., Vogel, H., Recht, L.D. and Sikic, B.I. (2005). High-resolution genome-wide mapping of genetic alterations in human glial brain tumors. *Cancer Res.* **65**, 4088–4096.

Chapter 9

Methylation Analysis by Microarray

Daniel E. Deatherage, Dustin Potter, Pearlly S. Yan, Tim H.-M. Huang, and Shili Lin

Abstract

Differential methylation hybridization (DMH) is a high-throughput DNA methylation screening tool that utilizes methylation-sensitive restriction enzymes to profile methylated fragments by hybridizing them to a CpG island microarray. This array contains probes spanning all the 27,800 islands annotated in the UCSC Genome Browser. Herein we describe a DMH protocol with clearly identified quality control points. In this manner, samples that are unlikely to provide good read-outs for differential methylation profiles between the test and the control samples will be identified and repeated with appropriate modifications. The step-by-step laboratory DMH protocol is described. In addition, we provide descriptions regarding DMH data analysis, including image quantification, background correction, and statistical procedures for both exploratory analysis and more formal inferences. Issues regarding quality control are addressed as well.

Key words: DNA methylation, differential methylation hybridization (DMH), CpG islands (CGI), microarray.

1. Introduction

The epigenome of a cell is a combination of important heritable characteristics that coordinate with DNA sequence information to modulate gene transcriptions. There are two general classes of epigenetic modifications: one that alters the residues on DNA-associated proteins (histones) and the other that adds methylation marks to the cytosine residues of CG dinucleotides. While the methylation state of DNA is often associated with particular

---

Daniel E. Deatherage and Dustin Potter have contributed equally to the work described in this chapter.

Jonathan R. Pollack (ed.), *Microarray Analysis of the Physical Genome: Methods and Protocols, vol. 556*
© Humana Press, a part of Springer Science+Business Media, LLC 2009
DOI 10.1007/978-1-60327-192-9_9 Springerprotocols.com

types of histone modifications *(1)*, this chapter focuses only on detecting DNA methylation in a genome-wide approach termed Differential methylation hybridization (DMH).

In the human genome, the occurrence of CG-dinucleotides is infrequent. They usually occur in clusters known as CpG islands or CGIs. There are different ways to annotate CGIs. Classically, if stretches of DNA longer than 500 bp have a total C and G greater than 55%, and the observed CG sites divided by the expected CG sites are greater than 65% *(2)*, this region is classified as a CGI. While the criteria for determining the presence of CGIs vary somewhat between research groups, it is of interest that 60–80% of the annotated islands occur around the promoter regions of known genes *(3)*.

In diseased states, such as cancer initiation and progression, DNA methylation in promoter CGIs is often associated with reduced expression or silencing of the genes involved *(4)*. Although other histone modifications and the recruitment of key factors to the promoters are also involved in this process, DNA methylation analyses are well established and can be readily conducted on large cohorts of patient samples and even in archival material.

The study of DNA methylation can be subdivided into two key types: targeted and genome-wide analyses. In the targeted approach the goal is to survey the methylation status of a selected genomic region with high resolution and specificity. However, in a genome-wide approach the goal is to capture multiple genomic regions which harbor DNA methylation. In this chapter, in additional to the global DMH analysis (*see* **Fig. 9.1**), we will also discuss using a targeted approach, MassARRAY/EpiTyper 1.0, for validating global targets (*see* **Fig. 9.2**).

Modern systems-biology technologies have dramatically altered the research landscape of biology by introducing the difficulties inherent in working with a high-dimensional data space where a single sample may be associated with tens of thousands of measurements. DMH-data are not free of these difficulties: a single array provides over 244,000 data points associated with more than 20,000 CGIs. On top of the difficulties inherent in high-dimensional data, the signal-to-noise ratio of the raw DMH data will decrease the sensitivity of standard statistical analysis methods and thus the data must be preprocessed appropriately in order to increase the signal-to-noise ratio. Our goal is not to provide a guide to DMH data analysis as this has been described before *(5)*; we instead describe the theoretical motivation behind the varying preprocessing and analytical methods available and provide a framework for deciding the approach most suitable for a given scientific enquiry.

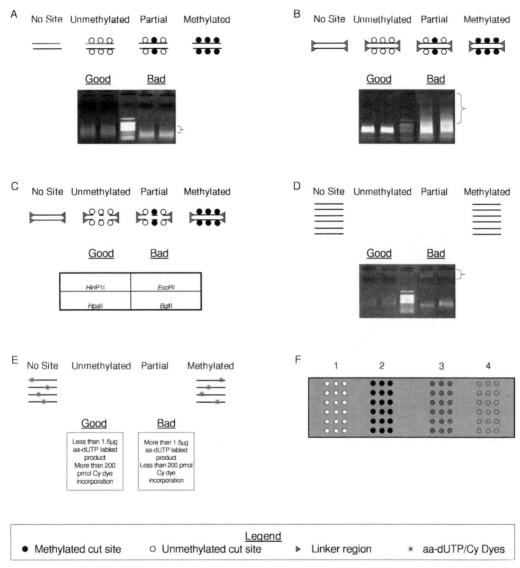

Fig. 9.1. DMH protocol outline. (**A**) **DNA sonication:** Genomic DNA is fragmented to ~500–800 bp in length. Gel shows typical DNA smears of good and bad samples. (**B**) **End-blunting and linker-ligation:** Sonicated DNA is end-repaired and used as a template for linker-adapters attachment. Gel shows a good sample (smear pattern similar to the sonicated DNA) and a bad sample (smear pattern has high-MW masses that are not seen in the sonicated DNA). (**C**) **Methylation-sensitive restriction:** The Linker ligated fragments are digested with any restriction enzyme which recognizes a sequence containing a CG di-nucleotide and is only able to digest the DNA if it lacks a methylated cytosine. (**D**) **Linker-mediated PCR:** The fragments which survive digestion are amplified using primers complementary to the linker sequence. As depicted, fragments containing unmethylated or partially methylated sites will not be amplified. (**E**) **Dye labeling:** Klenow fragments are used to incorporate amino-allyl dUTPs (aa-dUTPs) into the amplified fragments. These aa-dUTPs then serve as attachment sites for the Cy dyes. (**F**) **Microarray outcomes:** After the test sample and control sample are mixed at equal concentrations, they are allowed to hybridize to the microarray potentially giving any of the following outcomes for each probe: **1** (*pseudo-red*; empty circles) hypermethylation of test sample as compared to the control sample; **2** (*pseudo-green*; filled light gray circles) hypomethylation of test sample; **3** (*pseudo-yellow*; filled dark gray circles) equal methylation of both samples; **4** (no signal above background; filled black circles) no hybridization to the probes can be caused by poor interaction between the probe and fragment, or fragments not seen because they are unmethylated and digested.

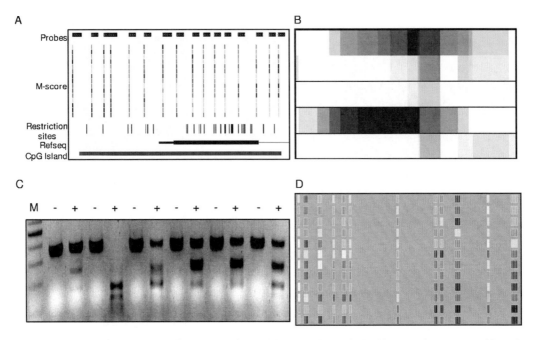

Fig. 9.2. DMH data validation. **(A) Identifying the region:** Bed files, which contain the *M*-score values generated from the microarray signals, are visualized using the UCSC genome browser. The advantage of using the genome browser is in the ease of determining associated genes and if any easily detectable patterns exist in the methylation of the probes. It is also useful to visualize the density of restriction enzymes which can be used for COBRA analysis. **(B) Smudge plots:** Smudge plot (when used in conjuncture with the genome browser) representing a visualization of the most methylated regions of all the test samples in the study. Primers for COBRA analysis can then be generated for the regions showing the most methylation in DMH (also containing the necessary restriction sites for downstream validation). **(C) COBRA analysis:** Agarose gel showing qualitatively the methylation state of a series of samples. Lower-molecular-weight bands seen in the digested lanes (+), compared to the mock digested lanes (−), signify the methylation of the restriction site prior to the bisulfite conversion reaction. **(D) Quantitative MassARRAY data:** MassARRAY analysis gives quantitative methylation values listed as percentage methylated from 0 to 100%. These values for each CpG unit can be adjusted with the use of a standard curve and a large sample set to give more accurate readings (described in **Section 3**). The figure represents the adjusted values with black being 100% methylated and white being nonmethylated.

2. Materials

2.1. Genomic DNA Isolation

1. QIAamp DNA Mini Kit (Qiagen, Valencia, CA).
2. ND-1000 Spectrophotometer (NanoDrop Technologies, Wilmington, DE).

2.2. DNA Fragmentation

1. Bioruptor 200 (Diagenode SA., Liege, Belgium).

2.3. DMH Amplicon Preparation

1. 100 mM dNTPs (Fisher Scientific).
2. 3 U/μL T4 DNA Polymerase (New England Biolabs (NEB), Ipswich, MA).

3. Zymo DNA Clean & Concentrator-5 columns (Zymo Research Corp., Orange, CA).

4. Linker sequences: JW103 (5′-GAA TTC AGA TC-3′) and JW102 (5′-GCG GTG ACC CGG GAG ATC TGC ATT C-3′).

5. PEG-6000 (Sigma-Aldrich, St. Louis, MO).

6. 400 U/μL T4 Ligase (NEB).

7. 2 U/μL DeepVentR (exo-) DNA Polymerase (NEB).

8. Methylation-sensitive restriction endonucleases: 10 U/μL *Hin*P1I (restriction site: 5′-G↓CGC-3′, NEB); 10 U/μL *Hpa*II (restriction site: 5′-C↓CGG-3′, NEB).

9. 2 U/μL Deep Vent (exo-) DNA polymerase (NEB).

10. Aminoallyl-dUTP (Fermentas, Glen Burnie, MD).

11. BioPrime labeling kit (Invitrogen, Carlsbad, CA).

12. Sodium carbonate.

13. 3 M Sodium acetate buffer solution (Sigma-Aldrich).

14. Cy-Dye Post-labeling reactive dye (Amersham Health Inc., Princeton, NJ).

2.4. Microarray Hybridization

2.4.1. Array Hybridization

1. Hybridization chamber (Agilent Technology).

2. Hybridization oven (Agilent Technology).

3. Gasket slides (Agilent Technology).

4. Human Cot-1 DNA (Invitrogen).

5. Oligo aCGH/ChIP-chip Hybridization Kit (Agilent Techology).

6. Human CpG Island ChIP-on-chip Microarray Set (Agilent Technology).

2.4.2. Array Washing

1. Slide rack.

2. Slide tank.

3. Rotating/rocking platform.

4. Stabilization and Drying Solution (Agilent Technology).

5. AccuGENE 20X SSPE buffer (Lonza, Rockford, ME).

6. Sarcosine (Fisher Scientific).

2.4.3. Array Visualization

1. GenePix Pro 6.0 (Molecular Devices Corp., Sunnyvale, CA).

2. GenePix 4000B Microarray Scanner (Molecular Devices Corp.).

2.5. Target Validation

1. EZ DNA Methylation kit (Zymo Research, Orange, CA).

2. 10X Amplitaq Gold polymerase, supplied with buffer and magnesium chloride solution (Applied Biosystems, Foster City, CA).

3. dNTPs.

4. DMSO.

5. Gene-specific primer pairs.

6. Methylation-sensitive restriction endonuclease.

3. Methods

3.1. Genomic DNA Isolation

1. A QIAamp DNA Mini Kit is used to obtain high-quality genomic DNA as per the manufacturer's direction.

2. As the ratio of DNA to linker is vital to achieving optimal linker-ligation efficiency, it is important to determine the concentration of the genomic DNA after isolation. DNA concentration and purity can be obtained rapidly using ND-1000 Spectrophotometer.

3. A 0.7% agarose gel should be used to determine the quality of the isolated DNA (see **Note 1**).

3.2. DNA Fragmentation and Restriction

1. Reduction in genome complexity can be achieved by using restriction enzymes that do not cut at CG-rich regions. This includes enzymes such as *Mse*I (T↓TAA), *Bfa*I (C↓TAG), *Nla*III (CATG↓), or *Tsp509*I (↓AATT). *See* **Note 2** for more information.

2. An alternative approach to simplify the genome is to fragment DNA by sonication.

3. There are different types of sonicators and we provide our workflow below as a general guideline. As many different factors affect DNA fragmentation, tight control of all aspects is needed to reproducibly generate the desired fragment size.

4. Pre-cool the Bioruptor with an ice–water mixture, and remove all ice before initial sonication step. Add a predetermined amount of ice back to the Bioruptor to bring the water level to the appropriate height before the first cycle of sonication.

5. The Bioruptor is operated in a cyclic manner rather than continuously running. We have found that eight on–off cycles of 30 each yields DNA fragments with average 500 bp. Ice should be replaced after every two cycles, and both the water and ice should be replaced after four cycles to reduce fluctuation of water temperature.

6. Gel electrophoresis (2% agarose gel) is used to determine the fragment size following the sonication. If the fragmentation pattern does not match the desired pattern, additional sonciation cycles can be performed.

7. *See* **Note 3** for additional information.

3.3. DNA Preparation and Amplification

1. End-repair 100–200 ng of sonicated DNA by adding the following:
   a. T4 DNA Polymerase (3 µL)
   b. 2 mM dNTPs (4 µL)
   c. 10X BSA (2.5 µL)
   d. 10X NEB buffer#2 (5 µL)
   e. Enough water to reach a final volume of 50 µL

2. Incubate at 37°C for 2.5 h. A Zymo DNA purification column is used to purify the product. Elute DNA with 29.5 µL water.

3. Linker adapters should be freshly prepared heating equal molar amounts of JW102 and JW103 oligonucleotides to 95–100°C for 5 min and allowing the mixture to cool to room temperature gradually. This is best accomplished by placing the tube of linkers in a beaker of boiling water for 5 min and then removing the beaker from the heat source, allowing the temperature to equilibrate to ambient temperature.

4. All reagents are kept on ice while the reaction mixture is prepared. To the 29.5 µL of end-repaired DNA from Step 1 add the following to give a final volume of 40 µL:
   a. Annealed linker adapters (2.5 µL)
   b. 10 mM ATP (1 µL)
   c. 50% PEG-6000 (2.5 µL)
   d. 10X T4 Ligase buffer (4 µL)
   e. T4 ligase (0.5 µL)

5. For best control, a thermocycler is used to incubate the reaction mixture at 14°C for 2 h.

6. The efficiency of the linker-ligation is evaluated by performing a test PCR. A volume of 1 µL of ligation product is used as a template in a PCR reaction:
   a. 10 µM JW102 (0.2 µL)
   b. 10 mM dNTP (0.4 µL)
   c. 10X ThermoPol Buffer (2 µL)
   d. DeepVent DNA polymerase (0.4 µL)
   e. Water (14 µL)

7. Amplification conditions are as follows:
   a. 55°C for 2 min, 72°C for 5 min, 95°C for 2 min.
   b. 55°C for 30 s and 72°C for 1 min (for 17 cycles).
   c. 55°C for 30 s and 72°C for 10 min.

8. Gel electrophoresis is used to visualize the efficiency of the ligation. After separation on a 1.5% agarose gel, one should evaluate smear patterns critically as follows: high-MW

smears/bands not previously seen after sonication are indicative of an overligated sample; an extremely faint smear or high-MW band which barely migrates beyond the well is indicative of too much material being lost in the preceding clean-up steps. Either of these scenarios indicates a failed sample and requires restarting the DMH protocol from the linker-ligation step by re-evaluating the concentration and the quality of the sonicated DNA.

9. If the ligation test PCR smear mimics the smear pattern of the post-sonication DNA, the sample passes this QC checkpoint and is ready to proceed to the next step. While there are many methylation-sensitive enzymes capable of interrogating the methylation status of the linker-ligated fragments, we currently use *Hpa*II and *Hin*P1I for this purpose. Two different enzymes are used to decrease the likelihood of incomplete digestion by any one enzyme and to reduce the possibility of false-positive results. After sequential digestions, the product is purified with a Zymo column and eluted in 40 µL water.

10. The purified restricted fragments will serve as templates for a final linker-mediated PCR. The PCR mix consists of an appropriate amount of template, typically 4–10 µL restricted DNA is used (*see* **Note 4**).
    a. 10X ThermoPol Buffer (20 µL)
    b. 10 mM dNTP (4 µL)
    c. 10 µM JW102 (2 µL)
    d. DeepVent DNA polymerase (4 µL)
    e. enough water to bring the final volume to 200 µL

11. This mixture is divided into four PCR tubes (50 µL per tube) for the actual amplification. Amplification conditions are as follows:
    a. 55°C for 2 min, 72°C for 5 min, 95°C for 2 min.
    b. 55°C for 30 s and 72°C for 1 min (for 24 cycles).
    c. 55°C for 30 s and 72°C for 10 min.

12. A Qiaquick column is used to purify the combined PCR products. We typically see a yield of 0.8–1.5 µg DNA when eluting twice with 40 µL water. A 1.5% agarose gel should be used to assay the smear pattern of the PCR products. Again the pattern should mimic the initial smears.

3.4. Fluorescent Dye Labeling

1. We indirectly incorporate the fluorescent dyes (Cy 5 and Cy 3) into the PCR products through the coupling of the dye molecules to an intermediate moiety, aminoallyl dNTP. PCR products (600 ng each from test and control samples)

diluted in 68 μL water are combined with BioPrime 2.5X random primers (60 μL) and denatured at 95°C for 5 min and placed on ice for 3 min.

2. Once the mixture has cooled the following reagents are added, and the reaction is incubated at 37°C for 6 h products

   a. 10X dNTP (2 mM dATP, dCTP and dGTP, 0.35 mM dTTP) (15 μL)

   b. 10 mM aminoallyl-dUTP (4 μL)

   c. Klenow (40 U/μl) (3 μL)

3. The reaction is purified using a Qiaquick column and eluting twice with 40μL water. We typically see a yield of between 6 and 10 μg of DNA. The products are then dried using a savant system and re-suspended in 3 μL of water.

4. The nature of the dyes requires that the fluorescent dye coupling step be carried out in the absence of direct fluorescent lighting. Before being combined with the test sample, 3 μL 0.1 M Na carbonate buffer (pH 9.0) is used to reconstitute the Cy 5 dye. The 3 μL of re-suspended sample should be added to the Cy 5 dye and should be mixed every 30 min for 3.5 h. The Cy 3 dye is similarly reconstituted before the control sample DNA is added. The Cy 3 control sample is only mixed for 1.5 h as the incorporation of the Cy 3 dye is much more robust than the Cy 5 dye. It is important to stagger the starting time of these mixtures so they are ready at the same time. Add 100 mM sodium acetate (pH 5.2) mixed 1:1 with water (70 μL) to each reaction mixture at the end of the incubation period.

5. After purifying the dye mixtures with Qiaquick columns, labeled samples free of unincorporated dyes are eluted with 80 μL water.

6. The following absorbance readings should be obtained to determine the concentration of DNA and incorporation of each dye: DNA concentration (260 nm), Cy3 incorporation (550 nm), and Cy5 incorporation (650 nm).

3.5. Microarray Hybridization and Washing

1. Mix Cy5- and Cy3-labeled samples such that an equivalent amount of 300 pmol of each fluorescent dye is present in each hybridization mix together with the following reagents:

   a. Cot-1 DNA (20 μg)

   b. Agilent blocking buffer (50 μL)

   c. Agilent hybridization buffer (250 μL)

   d. Enough water to bring the total volume to 500 μL

2. Denature the resultant mix for 3 min at 95°C, and then incubate for 30 min at 40°C.

3. It is important that the hybridization equipment (hybridization chamber base, chamber top, clamp, chamber thumbscrew, hybridization gasket slide, and CpG island microarray) is assembled in a location close to the hybridization oven to permit an uninterrupted workflow.

4. Place the gasket slide in the chamber base and carefully add the labeled samples to the gasket slide. Place the microarray slide on the gasket slide with the printed side face down. Position the chamber top on top of the slide assembly, and use the chamber thumbscrew and clamp to seal the hybridization chamber. It is important that the chamber thumbscrew is only turned 90° past snug.

5. The entire chamber is then placed in a rotating hybridization oven set at 65°C for 16–20 h rotating at a speed of 10.

6. In order to effectively eliminate un-hybridized or cross-hybridized probes from targets, the Agilent Stabilization and Drying Solution should be pre-warmed at 37°C. A 1-L solution containing 299 mL 20X SSPE buffer and 1 mL of 5% sarcosine should be prepared freshly.

7. Disassemble the hybridization chamber and place the gasket slide in the 1 L SSPE-sarcosine buffer which was freshly made. Carefully pry open the gasket slide and move the hybridized microarray directly into a slide rack sitting in the same buffer. Wash by allowing the rack to gently rock for 5 min.

8. Transfer the slide rack to a 1 L solution containing 3 mL 20X SSPE buffer and wash for an additional 5 min. The slide rack is then moved to the pre-warmed Stabilization and Drying Solution and allow to gently rock for 1 min.

9. Using a slow and controlled motion, pull the slide out of the Stabilization and Drying Solution and place in a light-protected slide box. The dried microarray should be scanned immediately.

3.6. Image Quantification

1. The oligonucleotides (45–60 mers) designed to span human CpG islands are printed onto glass slides using Agilent's 60-mer SurePrint technology. As the SurePrint technology utilizes a non-contact inkjet approach to generate the targets, defects due to surface tension interactions and print-tip variability will be a non-issue in this platform.

2. The CpG Island microarray contains nearly 244 K targets with 237,220 of these falling within 95 bp of a CpG island. The remainder of the targets are designed for slide alignment, array quality assessment, or signal pre-processing.

3. The probes are uniformly distributed on the array within a 267 × 912 grid.

4. Several scanners are available for capturing hybridized micro-array signals. Scanners such as the Agilent DNA microarray scanner require little user input. The Axon scanner (GenePix Pro 6.0 software) requires user input to identify spot location and capture quantitative values for each spot on the array. Below are brief outlines regarding the operation of an Axon scanner:

   a. Although the program is fully automated, the scanning process should proceed with a set of pre-determination criteria to permit consistent scanned results.

   b. It is expected that only a small number of probes on the array should demonstrate differential signal between the Cy3 and Cy5 channel; therefore, the photo multiplier tube (PMT) settings are adjusted so that the overall distribution of the two channels is equivalent.

   c. These adjustments are implemented manually by the operator:

      i.   To reduce signal bleaching by repeat exposure to the scanning laser, we suggest the pre-scanning process should be confined to the top 25% of the array with continual fine adjustment of the PMT settings to arrive at intensity histogram plots depicting an overall balanced distribution of the two channels.

      ii. The full scan will be performed with the adjusted PMT levels.

      iii. Because the microarray pre-scan is performed on a small section of the microarray, the choice of PMT settings may be too intense resulting in signal saturations for either or both channels when the entire array is scanned. The objective is to generate scanned signals that span the entire dynamic range without resulting in signal saturation on the spots with high level of hybridization.

   d. The dynamic range of the scanner is between 0 and 2^{16} units. Thus, if the PMT settings are too high, quantitative values for probes with high intensity may be compromised because their signal is greater than the maximal scanning threshold.

   e. GenePix Pro 6.0 software will automatically determine spot size and location as well as signal intensity for the Cy3 channel, the Cy5 channels, and the background. The algorithm for determining spot location and size is highly accurate but not perfect. For example, dye blobs located close to target print sites would be reported as the hybridization signals. Therefore, it is important for the operator to scroll through the gridded microarray image to manu--ally flag mis-calls described above.

3.7. Background Correction

1. Signal intensity for a given probe is due to fluorescent signals from labeled DNA probes (true complementary hybridization to the DNA targets) as well as various background signals such as:

   a. Fluorescent signals from DNA probes that cross-hybridize non-specifically to the arrayed targets.

   b. Incomplete removal of labeled DNA cross-reacting with the slide matrices during the washing step.

2. The scanning software provides an intensity value for background signal that is the summation of fluorescent intensities from microarray substrate, labeled DNA that cross-reacts with the substrate and not the considered probe target, labeled DNA fragments that bled over from neighboring probes, and the occasional dye blobs.

3. The negative control probes provide a means for assessing the level of non-specific binding occurring on the array or within the neighborhood of a given probe.

4. There is no consensus within the microarray community with regards to the appropriate strategy for correcting for background noise. In our experience, background correction will introduce noise even as it corrects for background signals between and within experiments. Therefore, computational scientists have to work with biologists to arrive at a balanced approach for data pre-processing. Points to consider:

   a. If the objective is to identify differentially methylated probes from many similarly methylated but noisy probes, then the need to control for noise outweighs the need for background correction. A simple approach in this scenario will be the removal of probes with signal below background from further analysis.

   b. If the objective is to evaluate methylation differences between probes of interest, then an accurate signal intensity is needed and appropriate background corrections should be applied to probe intensities prior to comparison. One approach to correct for both cross-hybridization and substrate bleed-through would be to subtract a weighted average of the local background signal and the signals obtained from negative controls situated close to the probe of interest.

3.8. Quality Control

1. The CpG Island array has approximately 5,000 control probes dispersed evenly across the entire hybridization surface. These probes are designed for: image orientation, quality assessment of sample hybridization, measurement of background signals, and data normalization. *See* **Table 9.1** for detailed description of the control probes.

Table 9.1
Description of control probes

Name	Count	Purpose	Description
BrightCorner (HsCGHBrightCorner)	1 seq. rep. 14×	Used for slide orientation	Endogenous sequence with predicted high signal
DarkCorner	1 seq. rep. 35×	Used for slide orientation	Probe forms a hairpin and does not hybridize with sample
Structural negative (NegativeControl)	1 seq. rep. 675×	Measure of local background signal	Probe forms a hairpin and does not hybridize with sample
Biological negatives (e.g., NC1_00000002)	98 seq. rep. 6× 1 seq. rep. 108×	Measure of cross-hybridization	Random sequence that do not hybridize well to any sample
Reserve negatives (e.g., SM_01)	12 seq. rep. 40×		
Biological positives (PC_00000004)	1 seq. rep. 480×	Positive control	Endogenous sequence with predicted high signal
Deletion stringency probes (e.g., DCP_008001.0)	50 seq. rep. 2×	Positive control as well as assessment of mismatch effect	See* below
Intensity curve probes (e.g., LACC:SRN_800001, LACC:Intensity3, LACC:GD13C_10_1)	3,564 seq. rep. 1× 20 seq. rep. 12×	Signal normalization	Predicted to span the signal space using in-house models

* 10 probes, predicted to perform well by in silico analysis, are chosen randomly from a tiling database. In addition, four variants of these probes are printed on the array: a 1-bp deletion; a 3-bp deletion; a 5-bp deletion; and a 7-bp deletion. Deleted bases are chosen at random from the center of the probe sequence. The number after the "." indicates the number of bases deleted (e.g., DCP_008001.0 and DCP_008001.3 are from the same parent sequence and have 0- and 3-bp deletions, respectively).

2. Negative controls
   a. *Arabidopsis* control spots can be used for spike-in experiments or can be used as negative control spots.
   b. In the event when *Arabidopsis* fragments were not used as spike-in, we should see low signal intensities in the *Arabidopsis* control and the structural control spots.
   c. Probes with signal near or below the negative controls cannot be estimated reliably (even if signal is greater than local background)

3. Positive controls
   a. Many of the positive control probes are printed multiple times across the slide. These positive controls are determined empirically to be present in high abundance in many sample types.

   b. Some of the positive probes are from genomic regions (a.k.a. gene desert regions) known to contain few methyl-sensitive restriction cut sites.

   c. To gain some perspectives regarding the spatial variabilities in each hybridization experiment, one can track the signal variations of each type of positive controls that are arrayed multiple times across the slide. We expect to see these probes having similar signal intensities across the array.

3.9. Data Preprocessing

3.9.1. Data Cleaning

1. Most image quantification programs flag spots that do not pass internal QC criteria; these spots should be removed from subsequent analysis.

2. Flagging probes that fall below pre-determined thresholds for potential dismissal:
   a. Threshold for signal-to-noise ratio

   b. Threshold for the percent of foreground pixels with signal larger than background

   c. Threshold for the summarized signal

3. Determining thresholds:
   a. Criterion for threshold determination should be customized.

   b. The expected distribution should be derived from the actual distribution of relevant parameters across the chips.

   c. Determine what "normal" values should be and discard or down-weight probes whose signal is significantly outside of the "norm".

4. Composite scores may also be useful *(6)*

5. If it is desirable to have values for every probe, missing values may be imputed by a number of standard approaches:
   a. K-nearest neighbor *(7)*

   b. Single-value decomposition *(2)*

   c. Probe neighbor average

   d. The array is designed to tile CpG islands; hence, a probe with missing value will often be flanked by probes with signal above the established threshold. The average or median of a probe's "neighbors" can be used to impute the missing value.

3.9.2. Data Transformation

1. The differences between the two samples/channels are often reported as ratios. The compression of ratios between 0 and 1 may be problematic for downstream data analysis.

2. Log$_2$ transformation is often used to transform signal intensities prior to expressing them as fold changes. It is important to note that raw values below $2^{-16} \sim 0.00001$ are not considered and that the raw ratios below 1 are mapped to values between -16 and 0.

3. Notations used for log$_2$ transformation are as follows:
    a. *M* is denoted as the log-ratio as *M* is the pneumonic for "Minus" whereby $\log_2(Cy5/Cy3)$ is equivalent to $\log_2(Cy5) - \log_2(Cy3)$.

    b. *A* is denoted as the log-average as *A* is the pneumonic for "Add" whereby $A = 0.5*(\log_2(Cy5) + \log_2(Cy3))$.

3.9.3. Intra-slide Normalization Adjustment for Non-biological Differences Between the Two Channels

1. $M' = (M - L)/S$
    a. *L* is the mean or median log-ratio over a subset of probes *(8)*.

    b. Local weighted loess regression *(9, 10)* where the values of *A* are binned and a linear polynomial is fit to the binned data. *L* is smoothed at the boundaries of the bins so that the function is continuous in *A*.

    c. Robust linear regression *(11, 12)* where *L* is a linear polynomial in terms of *A* across multiple slides and replicates.

    d. *S* is a robust estimate of the scale such as the median absolute deviation or the loess regression of the absolute mean-normalized log-ratio on *A* *(13)*.

2. A spatial plot of the *M* values can often reveal the need for intra-slide normalization as well as which normalization procedure to employ.
    a. *MA*-plots are two-dimensional scatter plots, plotting the relationship between *M* and *A*.
        i. There should be no discernable pattern relating *M* to *A*.

        ii. The expected value of *M* is zero and thus the plot should be centered at zero.

    b. Plotting *M*-values:
        i. Convert the quantitative values of *M* into a color intensity. Two common approaches are as follows:
            I. Set color value to be green for values below -15, red for values above 15, and a continuous color gradient for values in between. This color scheme is useful for detecting if dye abundance is correlated with spatial location.

II. Set color value to be blue for the probe with the lowest M value, yellow for the probe with the largest M value, and a continuous color gradient for values in between. This color scheme is useful for detecting a correlation between relative ranking of M-values and spot location.

ii. Plot the colors for each probe on a two-dimensional plot where the $x-y$ coordinate is associated with the location of the probe on the array.

iii. The resultant plot should have no discernable pattern.

3.9.4. Inter-slide Normalization Adjusts for Non-biological Effects Between Arrays

1. M-values should be scaled so that they have the same median-deviation across arrays.

2. Quantile normalization *(14)*:

   a. A transformation that brings the mean (median) intensity of all the arrays to the same level.

   b. If a common reference sample has been labeled with one of the florescent dyes, e.g., Cy3, then quantile normalization should be applied to this channel. The method for adjusting the other channel depends on the intra-slide normalization conducted, but should be adjusted in a manner to not alter the normalized log-ratio within the studied array.

   c. If a reference sample is not used, then one can use quantile normalization to transform the M values.

3.9.5. Adjustments for Probe Composition and Target Region Effects

1. The pre-processing methods described above do not consider/incorporate known biases in the assays. It is possible to correct for these effects using simple linear regression models *(15, 16)*.

3.10. Data Analysis

1. This approach is most useful for uncovering possible relationships among different samples within a study.

3.10.1. Exploratory Analysis (i.e., Clustering)

2. Data reduction:

   a. Most clustering methods will be overwhelmed by the noise in the data if the entire data set is used. To circumvent that:

   i. A probe flagged in any chip should be removed from consideration.

   ii. Only probes with high variance across arrays should be considered, as probes with lower variance will not have the power necessary to distinguish between traits.

   iii. Do not pre-select probes that distinguish between treatments as this will bias your analysis.

3. Metrics

   a. Most clustering procedures require the operator to select a distance function between the observed data points. Different metrics will likely produce different clusters.

   b. Euclidean (L_2) *(17)*

      i. Very common and easy to understand, though hard to interpret in the setting of high-dimensional probe intensity data.

   c. Manhattan (L_1) *(18)*

      i. Also common and easy to understand but difficult to interpret.

   d. One minus absolute correlation *(19)*

      i. Highly correlated points will be closer than uncorrelated points.

   e. Mutual information (MI) *(20)*

      i. The MI between two random variables $\mathbf{X}$ and $\mathbf{Y}$ is given by:

$$I(\mathbf{X},\mathbf{Y}) = \sum_i \sum_j p_{ij} \left(\log(p_{ij}) - \log(p_i p_j)\right)$$

         where p_i and p_j is the probability that $\mathbf{X} = x_i$ and $\mathbf{Y} = y_i$, respectively.

      ii. The values for p can be estimated from the data.

      iii. MI is a generalized measure of correlation since the distance is zero if and only if $\mathbf{X}$ and $\mathbf{Y}$ are statistically independent.

3.10.2. Detecting Regions of Significantly Differentiated Methylation

1. A direct approach is to utilize threshold value to make a call of significance (e.g., probes with M values (i.e., log ratio) above 1 or below -1 are differentially methylated).

   a. One of the drawbacks for such a method is the inability to assess the methylation calls statistically. This method also cannot incorporate information derived from neighboring probes (an expected trend of co-methylation in nearby genomic regions).

2. *M*-score *(5)*:

   a. A simple kernel smoothing function termed *M*-score is used to integrate probe-level information within a sliding window to portray regional methylation events.

   b. As an example, the *M*-score of each probe with respect to other probes within 1-kb region of the genome (500-bp upstream and 500-bp downstream) is calculated as follows:

      i. Probes are ranked according to their normalized log ratios.

      ii. An arbitrary cutoff, *n*, is set (e.g., the top and bottom 25th percentile).

iii. M-score = (#probe$_{\text{log upper } n^{th}}$ − #probe$_{\text{log lower } n^{th}}$) / total probes in 1-kb window)

3. Parametric tests for discovering differential methylation.
   a. t-test (p-values) *(21)* and ANOVA (F-statistic) *(22)*
      i. Both tests can be used to discover loci with the power to independently differentiate between cases.
      ii. Both tests are sensitive to outliers
   b. Significance analysis of Microarrays (SAM) *(23)*
      i. A method that scores each probe intensity with respect to the change in intensity relative to the standard deviation of repeated measurements.
      ii. Significance of probes with score greater than a threshold is determined via a permutation test.

4. Non-parametric tests for discovering differential methylation.
   a. Wilcoxon signed-rank *(24)*
      i. Alternative to the t-test for discovering loci that individually differentiate between two groups.
      ii. Estimation of p-values assumes symmetry of distribution (may not be supported by the data)
   b. Peak detection *(25)*
      i. Model-based computational method for locating and testing peaks in landscape data generated using the M-score approach.
      ii. The methods proposed in *(20)* can be easily adapted to the DMH protocol. It is, however, important to consider hyper- and hypo-methylation events independently.

5. Permutation tests
   a. Often times the data will not satisfy the theoretical hypothesis for a given statistical test. Permutation tests will allow one to estimate the empirical distribution of the test.
   b. Choose a test statistic.
   c. Compute the test.
   d. Permute labels on samples at random and repeat Steps ii and iii above.
   e. Compute the number of cases in which the test statistic from the random sample is less than the test statistic from the real data.

6. False discovery rates and post-hoc p-values correction:
   a. For a given DMH experiment, large numbers of comparisons are made resulting in a high probability for false-positives. Therefore, the resultant p-values should be adjusted to correct for multiple testing.

b. Bonferroni and other similar methods
i. Too conservative due to correlation of test statistics *(26)*.
c. Westfall–Young correction *(27)*.
d. Baysian approaches.

3.11. Target Validation

1. Regions of the genome which are determined to be differentially methylated by *M*-score (uploaded as a .bed file) are visualized on the UCSC genome browser (http://genome.ucsc.edu) to determine the potential interest or importance of the region. Promoter CpG islands near tumor suppressor genes, transcription factors, or genes shown to be methylated in other tumor types are validated using a qualitative followed by a quantitative method.

2. Non-degenerate primers are designed against the bisulfite-converted DNA sequence to amplify the region in or around the probes identified as being differentially methylated in DMH analysis. It is important that the amplified region should contain at least one restriction site for an enzyme which has a CG-dinucleotide in its recognition site (e.g., *Bst*UI) as this site will be preserved in a methylated allele thereby providing restricted fragment(s) as a read-out of the hypermethylation status. As bisulfite-converted DNA will be used, it is important to adjust the PCR primers so that the amplified regions will be between 350 and 500 bp for optimal analysis of fresh-frozen samples and 100 and 150 bp for archival materials. If longer regions are desired, multiple primer sets can be designed to extend the interrogation area.

3. Samples of interest are bisulfite converted using the EZ DNA Methylation kit following the manufacturer's instructions. We, however, elect to elute the purified products with $2 \times 50\,\mu L$ of water and allow a 5-min incubation period before each spinning. The bisulfite-converted DNA will be used as templates in COBRA (Combined Bisulfite Restriction Analysis) assay.

4. Using $2\,\mu L$ of the bisulfite-converted DNA as templates, the following PCR reagents are added:
a. 10X Amplitaq Gold buffer ($2\,\mu L$)
b. 25 mM magnesium chloride ($2.4\,\mu L$)
c. 2.5 µM dNTP mix ($2\,\mu L$)
d. 10 µM Forward Reverse primer ($0.4\,\mu L$ each)
e. DMSO ($0.2\,\mu L$)
f. Water ($10.4\,\mu L$)
g. Amplitaq Gold polymerase ($0.2\,\mu L$)

5. PCR conditions will vary according to the optimal annealing temperature of the primers being used, but a typical amplification program is as follows:

    a. 95°C for 10 min.

    b. 95°C for 30 s, 58–62°C for 30 s, and 72°C for 1 min (for 45 cycles).

    c. 72°C for 10 min

6. Half of the PCR products (10 μL) is moved to a new tube containing the following reagents for COBRA digestion:

    a. Appropriate restriction enzyme (5 units)

    b. Matched restriction enzyme buffer (2 μL)

    c. Enough water to bring the total volume to 20 μL.

7. To minimize potential agarose gel artifacts, 8 μL water, and 2 μL restriction buffer should be added to the remaining 10 μL of PCR product. Both tubes are then incubated at the appropriate temperature for 1 h, and samples from both tubes will be run out side-by-side on a 1.5% agarose gel.

8. Presence of MW band(s) corresponding to the size of restricted fragment(s) in the restricted product lane is indicative of hypermethylation in the interrogated region as sodium bisulfite will abrogate any potential restriction sites if the CG sites are unmethylated. COBRA assay is a reliable and qualitative test to validate DMH results.

9. Often times, DMH analysis is performed on a subset of samples to identify regions of interest to be followed up in a large cohort of samples to derive statistical power. In this scenario, we will modify the COBRA primers to meet the specifications of MassARRAY assay (Sequenom, Inc.) to obtain quantitative methylation status of large number of samples in the region of interest.

10. While the MassARRAY assay is quantitatively accurate to within 5%, PCR amplification of bisulfite-converted DNA can introduce bias into the reaction by preferentially amplifying methylated or unmethylated species. In order to detect and correct for this potential bias, we use an artificial standard curve generated by combining 100% methylated DNA (CpGenome Universal Methylated DNA from Chemicon) and human blood DNA (isolated from whole blood using a standard QIAamp DNA Mini Kit) (*see* **Note 5**).

11. According to the manufacturer's guideline, 5′ modifications to the COBRA primers are added: forward primer modification, AGGAAGAGAG; reverse primer, CAGTAATACGACT CACTATAGGGAGAAGGCT.

12. The PCR conditions used in the COBRA assay are used to amplify the samples and standard curve samples. Experience tells us that the addition of the 5′ modifications does not significantly influence the PCR efficiency, but it is important that 5 μL of PCR product be examined on a 1.5% agarose gel to verify the presence of a dominant band of the predicted size. Neither an additional band having a much lower intensity than the expected product nor unused primer bands interfere with the MassARRAY assay. Samples which fail this criterion can be re-amplified before being submitted for MassARRAY analysis. If any of the standard curve samples fail to meet this criterion, it must be re-amplified.

13. A volume of 5 μL of PCR product is loaded into a 384- or 96-well plate for submission to a core facility. It is important to check with the core facility you will be submitting your samples to for specifics on how they wish to have the samples submitted.

14. When the data are returned to you, the excel worksheet that contains a listing of the individual CG units methylation levels can be used to transform the data into less biased results. The standard curve samples can be used to perform a standard linear regression for each CG unit and the resultant regression curve used to perform a linear transformation on all samples (including standards). As human blood DNA does show some level of methylation at some CpG sites, it is possible that percentages of greater than 100 or less than 0 can be reported at this point.

15. If negative values or values greater than 100 are created, a second linear transformation is performed. The second linear transformation requires setting the smallest value detected for each CG unit to zero, and setting the highest value detected to 1. Together the two linear transformations account for PCR bias (the first linear transformation), and allow for values that make sense in a biological context (the second linear transformation).

4. Notes

1. Between 60 and 80 ng of the DNA loaded on the gel should be of a single high-molecular-weight band, with absence of smears (signifying DNA degradation) and a low-weight bands (signifying RNA contamination).

2. The following is information regarding the length of product one should expect after digesting with non-CG-containing enzymes.

Enzyme	Mean length of digested product	Median length of digested product
*Mse*I	*158 bp*	*81 bp*
*Bfa*I	*387 bp*	*246 bp*
*Nla*III	*219 bp*	*136 bp*
*Tsp509*I	*140 bp*	*77 bp*

3. It is of primary importance that the sonicated product be as uniform across samples as possible so as to minimize experimental variance between samples. Any samples that are unable to conform to one another should be restarted to ensure the highest quality of the results.

4. The actual amount of DNA is determined empirically as the sample methylation status and the fragment size will alter this amount. It is important that the 4–10 µl of DNA correspond to between 0.8 and 1.5 µg of DNA. Too little DNA and there is a risk that there will not be enough for downstream steps, yet too much and there is a risk of bias if the reaction reaches a plateau.

5. A six-point standard curve (0, 20, 40, 60, 80, and 100% methylated) is created by bisulfite-converting 500 ng of the 100% methylated DNA and blood DNA mixed together at the appropriate concentrations using the EZ DNA Methylation Kit from Zymo Research. By mixing the DNA prior to bisulfite-conversion it is also possible to detect differences in bisulfite-conversion efficiencies.

References

1. Fuks, F (2005) DNA methylation and histone modifications: teaming up to silence genes. *Current Opinions Genetics Development* **15**:490–495.

2. Takai, D and Jones, PA (2002) Comprehensive analysis of CpG islands in human chromosomes 21 and 22. *Proceedings of the National Academy of Sciences* **99**: 3740–3745.

3. Davuluri, RV, Grosse, I, and Zhang, MQ (2001) Computational identification of promoters and first exons in the human genome. *Nature Genetics* **29**:412–417.

4. Jones, PA and Baylin, SB (2007) The epigenomics of cancer. *Cell* **128**:683–692.

5. Yan, P, Potter, D, Deatherage, D, Lin, S, and Huang, TH-M (2008) Differential methylation hybridization: profiling DNA methylation in a high-density CpG island microarray. *Methods in Molecular Biology, DNA Methylation Protocols*, 2nd edition.

6. Fare, TL, Coffey, EM, Dai, H, He, YD, Kessler, DA, Kilian, KA, Koch, JE, LeProust, E, Marton, MJ, Meyer, MR, Stoughton, RB, Tokiwa, GY, and Wang, Y (2003) Effects of atmospheric ozone on

microarray data quality. *Analytical Chemistry, ASAP Article* 10.1021

7. Troyanskaya, O, Cantor, M, Sherlock, G, Brown, P, Hastie, T, Tibshirani, R, Botstein, D, and Altman, RB (2001) Missing value estimation methods for DNA microarrays. *Bioinformatics* **17**(6):520–525.

8. Smyth, GK, Yang, Y-H., and Speed, TP (2003) Statistical issues in microarray data analysis. *Methods in Molecular Biology* **224**:111–136.

9. Wolfinger, RD, Gibson, G, Wolfinger, ED, Bennett, L, Hamadeh, H, Bushel, P, Afshari, C, and Paules, RS (2001) Assessing gene significance from cDNA microarray expression data via mixed models. *Journal of Computational Biology* **8**:625–637.

10. Yang, YH, Dudoit, S, Luu, P, and Speed, TP (2001) Normalization for cDNA microarray data. In Bittner, ML, Chen, Y, Dorsel, AN, and Dougherty, ER (eds.), *Microarrays: Optical Technologies and Informatics*, Volume **4266** of Proceedings of SPIE, San Jose, CA.

11. Finkelstein, DB, Gollub, J, Ewing, R, Sterky, F, Somerville, S, and Cherry, JM (2001) Iterative linear regression by sector. In Lin, SM and Johnson, KF (eds.), *Methods of Microarray Data Analysis. Papers from CAMDA 2000*, Kluwer Academic, Boston, MA, pp. 57–68.

12. Kepler, TB, Crosby, L, and Morgan, KT (2001) Normalization and analysis of DNA microarray data by self-consistency and local regression. *Santa Fe Institute Working Paper*, Santa Fe, NM.

13. Dean, N and Raftery, AE (2005) Normal uniform mixture differential gene expression detection for cDNA microarrays. *BMC Bioinformatics* **6**:173.

14. Bolstad, BM, Irizarry RA, Astrand, M, and Speed, TP (2003) A comparison of normalization methods for high density oligonucleotide array data based on bias and variance. *Bioinformatics* **19**(2):185–193.

15. Wu Z, Irizarry RA, Gentleman R, Murillo FM, and Spencer F (2004) A model based background adjustment for oligonucleotide expression arrays. *Journal of the American Statistical Association* **99**:909–918.

16. Johnson, WE, Li, W, Meyer, CA, Gottardo, R, Carroll, JS, Brown, M, and Liu, XS (2006) Model-based analysis of tiling-arrays for ChIP-chip. *PNAS* **103**(33):12457–12462.

17. D'haeseleer, P, Liang, S, and Somogyi, R (2000) Genetic network inference: from co-expression clustering to reverse engineering. *Bioinformatics* **16**:707–726.

18. Kaufman, L and Rousseeuw, PJ (1990) *Finding Groups in Data: An Introduction to Cluster Analysis*. Wiley, New York.

19. Eisen, MB, Spellman, PT, Brown, PO, and Botstein, D (1998) Cluster analysis and display of genome wide expression patterns. *Proceedings of the National Academy of Sciences USA* **95**:14863–14868.

20. Priness, I, Maimon, O, and Ben-Galcorresponding, I (2007) Evaluation of gene-expression clustering via mutual information distance measure. *BMC Bioinformatics* **8**:111.

21. Snedecor, GW and Cochran, WG (1989) *Statistical Methods*, 8th edition. Iowa State University Press, Ames, IA.

22. Lindman, HR (1974) *Analysis of Variance in Complex Experimental Designs*. W. H. Freeman & Co., San Francisco, CA.

23. Tusher, VG, Tibshirani, R, and Chu, G (2001) Significance analysis of microarrays applied to the ionizing radiation response. *PNAS* **98**:5116–5121.

24. Wilcoxon, F (1945) Individual comparisons by ranking methods. *Biometrics* **1**:80–83.

25. Zheng, M, Barrera, LO, Ren, B, and Wu, YN (2007) ChIP-chip: data, model, and analysis. *Biometrics* **63**:787–796.

26. Cao, H, Kane, D, Narasimhan, S, Sunshine, M, Bussey, K, Kim, S, Shankavaram, UT, Zeeberg, B, and Weinstein, J. Microarray data analysis, http://discover.nci.nih.gov/microarrayAnalysis/Statistical.Tests.jsp

27. Westfall, PH and Young, SS (1993) *Resampling-Based Multiple Testing*. Wiley, New York.

Chapter 10

Methylation Analysis by DNA Immunoprecipitation (MeDIP)

Emily A. Vucic, Ian M. Wilson, Jennifer M. Campbell, and Wan L. Lam

Abstract

Alteration in epigenetic regulation of gene expression is a common event in human cancer and developmental disease. CpG island hypermethylation and consequent gene silencing is observed for many genes involved in a diverse range of functions and pathways that become deregulated in the disease state. Comparative profiling of the methylome is therefore useful in disease gene discovery. The ability to identify epigenetic alterations on a global scale is imperative to understanding the patterns of gene silencing that parallel disease progression. Methylated DNA immunoprecipitation (MeDIP) is a technique that isolates methylated DNA fragments by immunoprecipitating with 5′-methylcytosine-specific antibodies. The enriched methylated DNA can then be analyzed in a locus-specific manner using PCR assay or in a genome-wide fashion by comparative genomic hybridization against a sample without MeDIP enrichment. This article describes the detailed protocol for MeDIP and hybridization of MeDIP DNA to a whole-genome tiling path BAC array.

Key words: Epigenetics, DNA methylation, hypermethylation, hypomethylation, CpG islands, methylated DNA immunoprecipitation, epigenetic methods and technologies, array-based methylation analysis, MeDIP aCGH.

1. Introduction

1.1. Brief Overview of Methylation

Epigenomics refers to the genome-wide study of heritable yet reversible changes, which do not alter the DNA sequence itself. Aberrant epigenetic changes such as global DNA hypomethylation and focal DNA hypermethylation are found in almost all tumor types and for many developmental diseases (1–4). Global hypomethylation characteristically increases with age, and is linked to chromosomal instability in cancer and other diseases (5, 6). Specific patterns of hypermethylated DNA are associated with the transcriptional silencing of genes correlating with cancer

Jonathan R. Pollack (ed.), *Microarray Analysis of the Physical Genome: Methods and Protocols*, vol. 556
© Humana Press, a part of Springer Science+Business Media, LLC 2009
DOI 10.1007/978-1-60327-192-9_10 Springerprotocols.com

progression, prognosis, and treatment response, therefore representing promising biological markers *(6–8)*. Drugs that inhibit methylation are used both as a research tool, to assess reactivation of genes silenced in cancer by hypermethylation, and in the treatment of some hematological malignancies *(9, 10)*. The identification of epigenetic targets may serve as the basis for new therapeutics aimed at underlying disease biology.

The study of DNA methylation, especially in the integration of epigenetic data with genomic and expression profiles, will improve our ability to identify causal DNA events and their impact on gene expression and disease behavior. Towards this aim, analysis of isolated methylated DNA fragments by a technique described here called methylated DNA immunoprecipitation (MeDIP), previously described by Weber *et al.*, can be accomplished by a number of whole-genome profiling technologies facilitating the survey of differential methylation on a whole-genome scale and locus-specific resolution *(11, 12)*.

1.2. Brief Outline of Technique

The availability of antibodies specific to 5′-methylcytosine has enabled the enrichment of methylated fragments of DNA by immunoprecipitation (IP). DNA is sonicated and divided into two fractions; one of which will be used for the IP reaction and the other to be used as whole-genome reference material during comparative analysis, called the input (IN) *(11)*. The DNA sample is incubated with anti-5′-methylcytosine antibodies, and immunoprecipitated products are purified and then analyzed for enrichment of methylated DNA relative to the reference DNA. The MeDIP technique is schematically outlined in **Fig. 10.1**. IP DNA can be compared to IN DNA using a number of array CGH (aCGH) platforms including CpG island arrays, whole-genome BAC or oligonucleotide (oligo) arrays *(12)*.

1.3. Validation

The efficiency of IP may be tested using standard real time quantitative PCR (qPCR) techniques. Reliable assessment of MeDIP efficiency can be performed with only a few nanograms of IP DNA per PCR assay. Alternative methods of validation are bisulfite sequencing and methylation-specific PCR *(13–15)*. To test the enrichment at a given locus by qPCR it is necessary to construct two sets of primers. The control set should amplify a segment of DNA that is devoid of CpG dinucleotides. The other primer set should amplify the locus one wishes to test (e.g., an imprinted region such as H19). Standard rules for qPCR primer design apply. When optimizing the MeDIP protocol, it is possible to use repetitive DNA sequences, imprinted regions, or developmentally silenced genes as test targets to monitor the enrichment efficiency – so long as their methylation status is known *(11)*. PCR assays should be performed for both primer sets on both

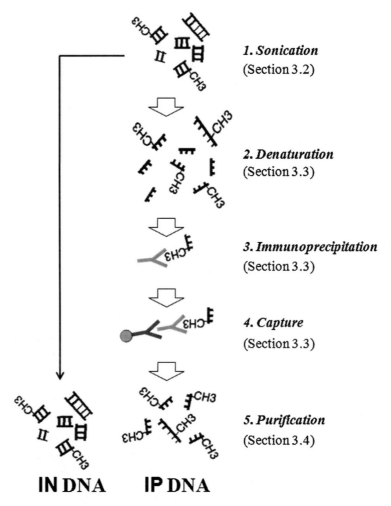

Fig. 10.1. **Methylated DNA immunoprecipitation (MeDIP).** A total of 1 µg of DNA is sonicated and divided into two tubes: one which will serve as a whole-genome reference for comparative analysis (IN DNA), the remainder of which will be immunoprecipitated (IP) (1). DNA for IP reaction is denatured (2), cooled, and incubated with primary antibody specific to methylated cytosine (3). A secondary antibody affixed with magnetic beads and specific to the primary antibody is added (4) and methylated DNA is thereby captured by use of a magnetic rack. IP reaction is then purified to remove proteins and buffer (5). IP DNA can then be compared to IN DNA for enrichment of methylation by PCR or array-based analysis.

the IN and IP fraction. Using threshold values for control and test primers, it is possible to calculate the fold enrichment using the $\Delta\Delta C_t$ method described by Livak et al. *(16)*.

1.4. Analysis of MeDIP DNA by Array CGH

1.4.1. BAC Array CGH

Analysis of MeDIP DNA by whole-genome BAC aCGH is a way to analyze DNA enriched for methylation in the sample on a whole-genome scale. Whole-genome methylation profiles can be obtained by differentially labeling IN and IP DNA from a single MeDIP reaction (or a pool of multiple MeDIP reactions for the same sample) with fluorescent nucleotide dyes, and then hybridizing to a whole-genome tiling path array such as the sub-megabase resolution tiling (SMRT) array CGH platform, which consists of ~27,000 overlapping BAC clones spanning the human genome *(11)*. These steps are outlined in **Fig. 10.2**. Dye ratios correlate to relative quantities of methylation-enriched IP DNA versus reference DNA (IN) bound to each BAC on the microarray.

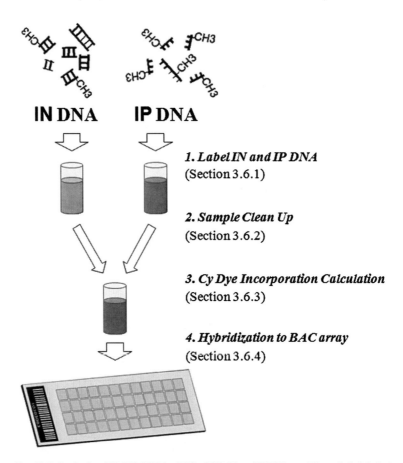

IN DNA   **IP DNA**

1. Label IN and IP DNA
(Section 3.6.1)

2. Sample Clean Up
(Section 3.6.2)

3. Cy Dye Incorporation Calculation
(Section 3.6.3)

4. Hybridization to BAC array
(Section 3.6.4)

Fig. 10.2. **Analysis of MeDIP DNA by BAC aCGH.** IN and IP DNA are differentially labeled with fluorescent Cy5 and Cy3 dyes, respectively (1). Unincorporated Cy dyes, buffers, and nucleotides are removed by use of exclusion columns (2). IN and IP reactions are combined, Cot-1 is added to bind repetitive DNA sequences and Cy dye incorporations determined by spectroscopy (3). Combined reactions are denatured. Cot-1 is allowed to anneal to labeled repetitive DNA and each sample is then applied to an array for hybridization (4).

1.4.2. Application of Other Array-Based Platforms to MeDIP Analysis

Similarly, other aCGH platforms including CpG island and genome-wide arrays can be used in conjunction with BAC aCGH or on their own. Examples include Agilent's Human Genome CGH and CpG island array (Agilent Technologies), and NimbleGen's Human Whole-Genome or custom-designed oligo arrays (NimbleGen Systems, Inc) (*see* **Note 1**).

Application of array-based platforms depends primarily on which platform will produce the most meaningful and reliable data in accordance with the experimenter's primary objective taking into account sample limitations. As previously described, one factor strongly influencing array alteration detection capacity is element size and distribution *(17)*. For methylation analysis, consideration of CpG island coverage and location (promoter or coding region for example) is especially relevant. The use of CpG island arrays in conjunction with whole-genome arrays may be most suitable for discovery-based research, whereas analysis of methylation status of CpG islands located in the promoters of target genes may be accomplished by lower-density CpG island or custom-designed oligo arrays. Sample IN requirement is another important consideration. Oligo platforms typically require 2 µg (Agilent) to 4 µg (NimbleGen) of product per dye channel (4–8 µg total); therefore, amplification of IP and IN DNA may be necessary (*see* **Note 2**). Since amplification may introduce bias, an alternative approach is to pool many MeDIP reactions to obtain the required amount.

2. Materials

2.1. DNA Preparation

1. Siliconized centrifuge tubes (1.7 ml SafeSeal Microcentrifuge Tubes, Sorenson BioScience)
2. Sterilized dH_2O

2.2. DNA Sonication

1. Sonicating device: Biorupter (Diagenode, UCD-200 TM)

2.3. Immunoprecipitation

1. IP Buffer: 10 mM sodium phosphate pH 7.0, 140 mM NaCl, 0.05% Triton X-100. Stored at room temperature.
2. Primary antibody: Anti-5′-methylcytosine Mouse mAb (162 33 D3) (CalBiochem). Stored at −20°C. Avoid freeze–thaw cycles.
3. Secondary antibody: Dynabeads M-280 Sheep anti-Mouse IgG (Dynal Biotech, Invitrogen). Stored at 4°C.
4. Rotating tube rack
5. Magnetic tube rack: magnetic stand (Invitrogen)

2.4. IP DNA Purification

1. 100% ethanol
2. 70% ethanol stored at −20°C
3. 3 M sodium acetate (pH 5.2)
4. 1:1 phenol:chloroform (buffered to pH 7)
5. Sterilized dH_2O

2.5. Validation by qPCR

1. MeDIP DNA (IP and IN)
2. iQ SYBR Green Supermix (Biorad laboratories)
3. Primers: Example of sequences for enrichment testing of immunoprecipitated human DNA:

   H19_F 5′-GGCGTAATGGAATGCTTGA

   H19_R 5′-CCTCGCCTAGTCTGGAAGC

   Producing a 63 bp product.

   CTRL_F 5′-GGTTCAGTTTATTGTCCTAAAATCAG

   CTRL_R 5′-TCAGCCAGACCAAAGCAAAT

   Producing a 92 bp product.
4. Real-time qPCR machine
5. Real-time PCR strip caps or real-time PCR plates and sealing tape

2.6. Analysis of MeDIP DNA by BAC aCGH

2.6.1. Labeling IN and IP DNA

1. Random primers buffer (PB) (5X Promega Klenow buffer and 7 µg/µl random octamers)
2. 10X dNTP mix (2 mM each dATP, dGTP, dTTP, 1.2 mM dCTP) (Promega)
3. Cyanine 5-dCTP (1 nmol/µl) (Amersham, GE Healthcare)
4. Cyanine 3-dCTP (1 nmol/µl) (Amersham, GE Healthcare)
5. Klenow fragment (9 U/µl) (Promega)

2.6.2. Sample Clean-Up

1. Microcon YM-30 column (Millipore)
2. Cot-1 DNA (1 µg/µl) (Invitrogen)
3. Hybridization buffer: DIG Easy Hyb (Roche Applied Science)
4. Sheared herring sperm DNA (20 mg/ml): Optional depending on slide chemistry (*see* **Note 3**)

2.6.3. Cy Dye Incorporation Calculation

1. ND-1000 Spectrophotometer V3.1.0 (NanoDrop Technologies Inc.)

2.6.4. Hybridization to BAC Array

1. Coverslips (22 × 60 mm, Fisher Scientific)
2. Hybridization cassette (Telechem)

2.6.5. Array Washing

1. Wash buffer: 0.1X SSC/0.1X SDS (pre warmed to 45°C)
2. Rinse buffer: 0.1X SSC (room temperature)
3. Coplin jar

2.6.6. Scanning

GenePix Professional 4200A (Molecular Devices)

3. Methods

3.1. Preparation of Samples

In one siliconized tube per sample, prepare 1 μg of DNA in 50 μl of sterilized dH$_2$O.

3.2. DNA Sonication

DNA sonication and MeDIP protocol must be performed in siliconzied tubes to prevent non-specific binding of proteins to tube walls. Here we describe a method to obtain 300–1,000 bp DNA fragments by sonication with an automated Biorupter (Diagenode, UCD-200 TM) (*see* **Note 4**).

1. Water in Biorupter must be at 4°C with layer of crushed ice on top to water mark.

2. Run for 5–7 min on automatic settings (30 s on 30 s off at maximum power).

3. Verify fragment size on 1% agarose gel with 100 bp ladder.

4. Remove 800 ng (40 μl) of sonicated product and place in siliconized 1.7 ml centrifuge tube for the IP reaction.

5. Set aside remainder (200 ng) to serve as IN reference DNA (store at 4°C).

3.3. Methylated DNA Immunoprecipitation

1. Denature the DNA that will be used for IP reaction (800 ng) at 95°C for 10 min in heat block.

2. Cool immediately on ice. Let DNA cool completely (~5 min on ice) before proceeding with next step.

3. Add 5 μg primary (anti-5′-methylcytosine) antibody.

4. Add IP buffer to a final volume of 500 μl.

5. Incubate a minimum of 2 h at 4°C in rotating tube holder (*see* **Note 5**).

6. Just before Step 5 is complete, prepare Dynabeads (coupled with secondary antibody) by washing. First resuspend the beads thoroughly in the vial by vortexing.

7. Transfer 30 μl (~ 2 × 10^7) of resuspended beads into a new siliconized tube. If performing more than one reaction, remove 30 μl of beads per reaction plus 1 (e.g., if doing eight reactions, remove enough beads for nine, i.e., 270 μl).

8. Place the tube on the magnetic rack for 2 min at room temperature.

9. Pipette off the supernatant. When pipetting off supernatant avoid touching the beads against inside wall (where the beads attract to the magnet) with the pipette tip.

10. Remove the tube from the magnet, and resuspend the beads in an excess volume of IP buffer (750–1,000 μl).

11. Repeat the wash once more, and then resuspend the washed beads in IP buffer in the original volume removed in Step 7.

12. Add 30 μl of washed Dynabeads to each IP reaction and incubate in a rotating tube holder for 2 h at 4°C.

13. After incubation with both antibodies is complete, place the tube on the magnetic rack for 2 min at room temperature.

14. Pipette off the supernatant. Avoid touching the inside wall of the tube (where the beads attract to the magnet) with the pipette tip.

15. Wash the bound Dynabeads three times in IP buffer, resuspending for the final time in 500 μl IP buffer.

16. Treat the reaction with 100 μg of proteinase K for 3 h at 50°C. Spike with 50 μg more proteinase K and continue to digest overnight at 50°C (*see* **Note 6**).

3.4. IP DNA Purification

1. Add 500 μl 1:1 phenol/chloroform pH 7 and vortex thoroughly.

2. Spin at 13,000 g for 10 min at room temperature.

3. Remove aqueous (top) fraction to a new tube (*see* **Note 7**).

4. Repeat Steps 1–3 one time to ensure complete removal of protein matter.

5. Add one-tenth volume 3 M sodium acetate (50 μl) and vortex.

6. Add 2 volumes (1,000 μl) of 100% ethanol; place in −20°C freezer for 20 min.

7. Spin at max speed at 4°C for 20 min.

8. Remove ethanol, pulse spin, and remove residual ethanol.

9. Add 500 μl cold 70% ethanol to wash. Vortex briefly, and spin at max speed for 20 min at 4°C.

10. Remove 70% ethanol, pulse spin, and remove residual ethanol by pipetting.

11. Dry pellets in heat block at 50°C for 10 min with caps open to remove all traces of residual ethanol.

12. Resuspend DNA pellet in sterile dH$_2$O (*see* **Note 8**).

13. Place tubes at 65°C for 10 min followed by 1 h incubation at 37°C to completely resuspend the DNA before use. Alternatively, after 37°C incubation store at −20°C for future use.

3.5. Validation by qPCR

1. Using iQ SYBR Green Supermix set up 25 μl PCR reactions with a final primer concentration of 0.6 μM for the forward and 0.6 μM for the reverse (*see* **Note 9**).

2. Real-time PCR should be performed using the following cycling parameters, with data collection occurring during the 60°C annealing/extension step:

   95°C for 5 min, (95°C for 15 s, 60°C for 45 s) × 40 cycles.

3.6. Analysis of MeDIP DNA by BAC aCGH

It is important to limit exposure of Cyanine dyes to light at all times.

3.6.1. Labeling IN and IP DNA

For every sample to be hybridized there will be two labeling reactions: one reaction tube for the reference (IN DNA) and one reaction tube for the IP DNA.

1. In sterilized PCR tubes, for every separate labeling reaction combine:

   a. DNA (50–400 ng)

   b. 5 μl of 5X PB buffer

   c. Dilute to 17 μl total volume with sterile dH$_2$O

2. Boil for 10 min at 100°C in a PCR machine.

3. Transfer immediately to ice for 2 min.

4. Vortex 10X dNTP thoroughly and add 4 μl to each labeling reaction.

5. Vortex and spin down Cyanine dyes. Add 2 nmol (2 μl of 1 nmol/μl stock) of Cyanine 5-dCTP (Cy5) to IN DNA and 2 nmol (2 μl of 1 nmol/μl stock) of Cyanine 3-dCTP (Cy3) to IP DNA.

6. Add 22.5 U Klenow (2.5 μl of 9 U/μl stock) to all reactions and mix gently by pipetting.

7. Incubate in the dark at 37°C overnight (~18 h).

3.6.2. Sample Clean-Up

It is important to limit the exposure of labeling reactions to light at all times. All steps in sample clean-up are performed at room temperature.

Using a Microcon YM-30 column (*see* **Note 10**):

1. Combine the two reactions (IN and IP) for each sample and add 100 μg Cot-1 DNA (1 μg/μl) to column (do not touch membrane with tip).

2. Spin at 13,000 g for 10 min in provided tubes.

3. Discard elution and add 200 μl sterile dH$_2$O to membrane and repeat spin in Step 2 to wash.

4. Discard collection tube and add 45 μl DIG Easy to column (*see* **Note 11**).

5. Leave at room temperature for 2 min then invert Microcon in a new tube and spin at 3,000 g for 3 min.

3.6.3. Cy Dye Incorporation Calculation

Remove 1.5 μl from the labeling reaction and measure Cy dye incorporation using NanoDrop Spectrophotometer (*see* **Note 12**). Use appropriate blank solution.

3.6.4. Hybridization to BAC Array

It is recommended to carry out all hybridization steps in a darkened room (*see* **Note 13**).

1. Denature labeled DNA (probes) at 85°C for 10 min.

2. Pulse spin, then place in 45°C incubator or heat block and allow Cot-1 to anneal to labeled repetitive DNA for 1 h.

3. Pre-warm a hybridization cassette to ~45°C.

4. Keeping the probe at 45°C in a heat block, pipette 45 μl of the probe solution onto the array slide (*see* **Note 14**).

5. Gently lower coverslip onto slide over probe solution avoiding bubbles.

6. Place the slide into a pre-warmed hybridization cassette and add 10 μl of H₂O in the lower groove (*see* **Note 15**).

7. Incubate for 36–40 h at 45°C.

3.6.5. Array Washing

All wash solutions should be at pH 7.0. All steps should be carried out in a darkened room (*see* **Note 13**).

1. Remove coverslip by gently submerging slide into a wash solution of 0.1X SSC/0.1X SDS until the coverslip falls off (*see* **Note 16**).

2. Wash the slides 2–3 times in 0.1X SSC/0.1X SDS pre-warmed to 45°C for 5 min with agitation in Coplin jar.

4. Rinse SDS by rinsing the slides 2–3 times in 0.1X SSC for 5 min with agitation until no bubbles appear.

5. Dry the slides with an air stream (oil free) or by centrifuging the slides in Falcon tubes at 900*g* for 3 min. (Blow off dust with air gun)

6. Store the slides in the dark (*see* **Note 17**).

3.6.6. Scanning

We currently use GenePix Professional 4200A (Molecular Devices). Slides are scanned at a 10 μm resolution, on automated optimization. The scanner requires a file with information about spot location to run automatic optimization (*see* **Note 18**).

Before exporting to visualization program, normalization programs are commonly used to correct for systematic variation inherent in experimental or technical processes such as gradients in the array images *(18)* (*see* **Note 19**).

4. Notes

1. Cost may be another consideration. Besides obvious differences in cost of different arrays and services, many arrays have only one spot per unique element and therefore two arrays may be required per sample if reproducibility calculations are desired.

2. WGA2 GenomePlex Complete Whole Genome Amplification (WGA) Kit (Sigma-Aldrich), or Bioscore Screening and Amplification Kit (Enzo Life Sciences) are most convenient.

3. Traditionally herring sperm is used to prevent non-specific background binding on slides coated with amine reactive groups. However, most current array slide chemistry (including the BAC aCGH protocol described herein) are comprised of non-reactive groups (aldehyde or epoxy) and therefore do not require the use of herring sperm.

4. DNA sonication is most easily and consistently accomplished in batch sets with an automated sonicator as described in this protocol. Sonication can also be performed using a conventional sonicating device such as Fisher Scientific Sonic Dismembrator (Fisher Scientific). Optimization of all sonicating devices must be performed for specific DNA quantity, total volume, and type and size of tube sonication will be performed in, prior to IP reaction. After initial optimization, base pair size should be periodically tested on an agarose gel over the course of normal usage.

5. Incubation with primary antibody can be left overnight.

6. It is normal for beads to clump during protein digestion. Overnight protein digestion is convenient but not necessary.

7. Tubes do not need to be siliconized from here on.

8. If hybridizing to BAC array then resuspend pellet in 13 μl dH$_2$O.

9. It is always necessary to determine primer efficiency separately. Each primer set will need an efficiency of near 100% for the $\Delta\Delta C_t$ method to be valid.

10. Microcon YM-30 columns filter species smaller than 50 bp of DNA, including all unincorporated Cy dyes, nucleotides, and buffers. Larger DNA fragments and undigested protein fragments will remain.

11. If slide chemistry requires, add an additional 4.5 μl (20 mg/ml) sheared herring sperm DNA. Please *see* **Note 3** for explanation.

12. Incorporations below 3.0 pmol/μl in either channel can produce variable results.

13. Preferably carry out all hybridization, washing, and scanning procedures in an ozone-controlled room. Ozone has been show to degrade Cy3 and Cy5 dye signal intensities *(19)*. Dedicating a room to hybridization procedures with an ozone monitoring and filtering device is ideal. Slides are particularly vulnerable when exposed to air, such as during washing, scanning, and storing.

14. Alternatively, place 45 μl probe solution onto coverslip and lower array slide onto coverslip.

15. This is to control humidity and prevent slide from drying out.

16. If the coverslip will not fall off, carefully remove by hand, trying not to scratch the array.

17. Humidity, light, and ozone may all contribute to signal degradation during storage.

18. Scan resolutions vary depending on array platform. For example, oligo arrays are usually scanned at 5 μm or may potentially require multiple scans per slide.

19. Because methylation data is not discrete data (i.e., many CpGs per element on array), segmentation analysis, such as applied to copy number analysis, is not as applicable. Generating lists of clones that are enriched or un-enriched for methylation can be accomplished by filtering normalized Cy3/Cy5 ratios based on thresholds. Ratios imported to software designed to view and analyze microarray data are very useful in comparing multiple samples or across assay types or platforms *(20, 21)*.

Acknowledgments

The authors wish to thank Bradley Coe, Chad Malloff, and Spencer Watson for useful discussion and assistance with this manuscript. This work was supported by funds from the Canadian Institutes for Health Research, Canadian Breast Cancer Research Alliance, Genome Canada/British Columbia, and National Institute of Dental and Craniofacial Research (NIDCR) grant R01 DE15965.

References

1. Esteller M, Herman JG. (2002) Cancer as an epigenetic disease: DNA methylation and chromatin alterations in human tumours. *J Pathol.* **196**, 1–7.

2. Feinberg AP. (2007) Phenotypic plasticity and the epigenetics of human disease. *Nature* **447**, 433–440.

3. Ozanne SE, Constancia M. (2007) Mechanisms of disease: the developmental origins of disease and the role of the epigenotype. *Nat Clin Pract Endocrinol Metab.* **3**, 539–546.

4. Shames DS, Minna JD, Gazdar AF. (2007) DNA methylation in health, disease, and cancer. *Curr Mol Med.* **7**, 85–102.

5. Jones PA, Baylin SB. (2002) The fundamental role of epigenetic events in cancer. *Nat Rev Genet.* **3**, 415–428.

6. Feinberg AP, Ohlsson R, Henikoff S. (2006) The epigenetic progenitor origin of human cancer. *Nat Rev Genet.* **7**, 21–33.

7. Glasspool RM, Teodoridis JM, Brown R. (2006) Epigenetics as a mechanism driving polygenic clinical drug resistance. *Br J Cancer.* **94**, 1087–1092.

8. Chekhun VF, Lukyanova NY, Kovalchuk O, Tryndyak VP, Pogribny IP. (2007) Epigenetic profiling of multidrug-resistant human MCF-7 breast adenocarcinoma cells reveals novel hyper- and hypomethylated targets. *Mol Cancer Ther.* **6**, 1089–1098.

9. Shames DS, Girard L, Gao B, Sato M, Lewis CM, Shivapurkar N, Jiang A, Perou CM, Kim YH, Pollack JR, Fong KM, Lam CL, et al. (2006) A genome-wide screen for promoter methylation in lung cancer identifies novel methylation markers for multiple malignancies. *PLoS Med.* **3**, e486.

10. Gryn J, Zeigler ZR, Shadduck RK, Lister J, Raymond JM, Sbeitan I, Srodes C, Meisner D, Evans C. (2002) Treatment of myelodysplastic syndromes with 5-azacytidine. *Leuk Res.* **26**, 893–897.

11. Weber M, Davies JJ, Wittig D, Oakeley EJ, Haase M, Lam WL, Schubeler D. (2005) Chromosome-wide and promoter-specific analyses identify sites of differential DNA methylation in normal and transformed human cells. *Nat Genet.* **37**, 853–862.

12. Wilson IM, Davies JJ, Weber M, Brown CJ, Alvarez CE, MacAulay C, Schubeler D, Lam WL. (2006) Epigenomics: mapping the methylome. *Cell Cycle* **5**, 155–158.

13. Frommer M, McDonald LE, Millar DS, Collis CM, Watt F, Grigg GW, Molloy PL, Paul CL. (1992) A genomic sequencing protocol that yields a positive display of 5-methylcytosine residues in individual DNA strands. *Proc Natl Acad Sci U S A* **89**, 1827–1831.

14. Gonzalgo ML, Liang G, Spruck CH, 3rd, Zingg JM, Rideout WM, 3rd, Jones PA. (1997) Identification and characterization of differentially methylated regions of genomic DNA by methylation-sensitive arbitrarily primed PCR. *Cancer Res.* **57**, 594–599.

15. Huang TH, Laux DE, Hamlin BC, Tran P, Tran H, Lubahn DB. (1997) Identification of DNA methylation markers for human breast carcinomas using the methylation-sensitive restriction fingerprinting technique. *Cancer Res.* **57**, 1030–1034.

16. Livak KJ, Schmittgen TD. (2001) Analysis of relative gene expression data using real-time quantitative PCR and the 2(-Delta Delta C(T)) Method. *Methods* **25**, 402–408.

17. Coe BP, Ylstra B, Carvalho B, Meijer GA, Macaulay C, Lam WL. (2007) Resolving the resolution of array CGH. *Genomics* **89**, 647–653.

18. Khojasteh M, Lam WL, Ward RK, MacAulay C. (2005) A stepwise framework for the normalization of array CGH data. *BMC Bioinformatics* **6**, 274.

19. Branham WS, Melvin CD, Han T, Desai VG, Moland CL, Scully AT, Fuscoe JC. (2007) Elimination of laboratory ozone leads to a dramatic improvement in the reproducibility of microarray gene expression measurements. *BMC Biotechnol.* **7**, 8.

20. Chi B, DeLeeuw RJ, Coe BP, MacAulay C, Lam WL. (2004) SeeGH – a software tool for visualization of whole genome array comparative genomic hybridization data. *BMC Bioinformatics.* **5**, 13.

21. Chari R, Lockwood WW, Coe BP, Chu A, Macey D, Thomson A, Davies JJ, MacAulay C, Lam WL. (2006) SIGMA: a system for integrative genomic microarray analysis of cancer genomes. *BMC Genomics* **7**, 324.

Chapter 11

Combining Chromatin Immunoprecipitation and Oligonucleotide Tiling Arrays (ChIP-Chip) for Functional Genomic Studies

Jérôme Eeckhoute, Mathieu Lupien, and Myles Brown

Abstract

Central to systems biology are genome-wide technologies and high-throughput experimental approaches. Completion of the sequencing of the human genome as well as those of a number of other higher eukaryotes now allows for the first time the mapping of all of the *cis*-regulatory regions of genes as well as the details of nucleosome position and modification. One approach to achieving this goal involves chromatin immunoprecipitation combined with DNA oligonucleotide tiling arrays (ChIP-chip). This allows for the identification of genomic regions bound by a given factor, its cistrome, or harboring a given epigenomic modification through hybridization on tiling arrays covering the entire genome or specific regions of interest. This technology offers an unbiased assessment of the potential biological function of any DNA associated factor or epigenomic mark. Through integration of ChIP-chip data with complementary genome-wide approaches including expression profiling, CGH and SNP arrays, novel paradigms of transcriptional regulation and chromatin structure are emerging.

Key words: Chromatin immunoprecipitation, tiling arrays, ChIP-chip, epigenetic, epigenomic, genomic, cistrome.

1. Introduction

A central component of chromatin immunoprecipitation on DNA tiling arrays (ChIP-chip) relates to the isolation of genomic regions bound by a particular factor or harboring a specific epigenomic mark from the whole genome of living cells. This crucial step is achieved through the covalent cross-linking of chromatin-bound

---

Jérôme Eeckhoute and Mathieu Lupien have contributed equally

Jonathan R. Pollack (ed.), *Microarray Analysis of the Physical Genome: Methods and Protocols, vol. 556*
© Humana Press, a part of Springer Science+Business Media, LLC 2009
DOI 10.1007/978-1-60327-192-9_11 Springerprotocols.com

proteins using formaldehyde alone or in combination with a protein–protein cross-linker such as disuccinimidyl glutarate (DSG) *(1, 2)*. Cells are then lysed and chromatin fragmented to produce a mixture of short pieces of chromatin still cross-linked to chromatin-bound factors. Subsequent immunoprecipitation using an antibody directed against the factor of interest allows for the selective isolation of chromatin regions bound by this factor in vivo. Following reverse cross-linking and elution, the limited amount of recovered DNA can be amplified and labeled for hybridization on DNA tiling arrays. These arrays can be designed to cover all the non-repetitive sequences of the whole-genome or specific regions of interest. They consist of contiguous probes of various lengths and spacing allowing for variation in the resolution of the analysis. Researchers routinely use pre-designed or custom-made tiling arrays that can be obtained through several companies whose platforms appear to perform equally well even if specificities exist *(3)*. After hybridization of the ChIPed DNA, arrays are scanned and a relative signal is ascribed to every single probe. Subsequent statistical analyses are used to determine regions bound by the factor of interest taking into account the signal given by several contiguous probes. A detailed discussion of ChIP-chip data analysis is provided in the subsequent chapter.

2. Material

2.1. Chromatin Immunoprecipitation (ChIP)

1. Protein A and/or G Dynal magnetic beads (Dynabeads).
2. ChIP dilution buffer: 1% triton, 2 mM EDTA, 150 mM NaCl, and 20 mM Tris-HCl pH 8.1.
3. Formaldehyde (Fisher).
4. Complete protease inhibitors caplets (Roche).
5. ChIP lysis buffer: 1% SDS, 10 mM EDTA, 50 mM Tris-HCl pH 8.1, and complete protease inhibitors cocktail (make up fresh every time).
6. RIPA buffer: 50 mM HEPES pH 7.6, 1 mM EDTA, 0.7% Na Deoxycholate, 1% NP-40, and 0.5 M LiCl.
7. Tris-EDTA (TE) buffer.
8. ChIP elution buffer: 1% SDS and 0.1 M NaHCO$_3$.
9. QIAquick PCR purification kit (Qiagen).
10. SybrGreen real-time PCR Master Mix (Applied Biosystems).

2.2. Preparation of the DNA for Hybridization onto the Tiling Arrays

1. PicoGreen DNA quantification reagent (Invitrogen).
2. RNase A (Sigma).
3. Proteinase K (Roche).

4. Phenol:chloroform:IAA (Ambion/Applied Biosystems).

5. Phase lock gels (Brinkmann Instrument Inc.).

6. Glycogen (Invitrogen).

7. End filling reagents: T4 DNA Pol Buffer (NEB buffer #2); BSA (NEB, 10 mg/ml); dNTP mix; T4 DNA pol (NEB, 3 U/μl).

8. Linker ligation reagents: 5X ligase buffer (Gibco) and T4 DNA ligase (NEB).

9. PCR reagents: 10X Thermopol buffer (NEB) and Amplitaq (Applied Biosystems).

10. Fragmentation reagents: DNAseI fragmentation mix (Affymetrix) and NEB buffer #4.

11. Labeling reagents: TdT buffer (Promega), TdT (30 U/μl, Promega), and Biotin (1 mM, Enzo Life Sciences).

3. Method

Preparation of DNA for ChIP-chip is detailed here as we have performed it using human cell lines and Affymetrix human tiling arrays 2.0 *(4–7)*. The protocol described here is subdivided into two sections and notes indicate alternative approaches to the described protocol. The first section describes the different steps involved in the initial ChIP procedure. The second section deals with DNA amplification by ligation-mediated PCR (LM-PCR) (*see* **Note 1**), its fragmentation and labeling for subsequent hybridization on arrays.

3.1. Chromatin Immunoprecipitation (ChIP)

1. Prepare antibody-Dynal magnetic beads to be used for the immunoprecipitation (*see* **Note 2**). For each sample, wash 20 μl of A or G Dynal beads (a mixture of both can also be used with some antibodies) three times with 500 μl of cold PBS plus BSA (5 mg/ml) using a magnetic concentrator. Incubate beads with your antibody of choice in cold PBS plus BSA (5 mg/ml) for at least 6 h in 500 μl of PBS plus BSA (5 mg/ml) on the mixer at 4°C. Typically 4 μg of antibody per ChIP is sufficient (*see* **Note 3**). Then, wash beads twice with cold PBS plus BSA (5 mg/ml), resuspend in ChIP dilution buffer, and add to chromatin at Step 11.

2. Prepare a PBS solution containing 1% formaldehyde and warm to 37°C (*see* **Note 4**).

3. Take the cells out of the incubator and quickly replace the culture media with the warmed PBS solution containing 1% formaldehyde and put the cells back in the incubator for

10 min. At that time, the cells need to be in the conditions (confluence, treatment...) that the investigator wants to investigate (*see* **Note 5**).

4. Remove the formaldehyde-containing PBS and quickly wash the cells once with cold PBS plus BSA (5 mg/ml) and once with cold PBS.

5. Harvest the cells by scraping into 350 or 500 µl ice-cold PBS complete (PBS containing protease inhibitors) for a 10 cm or a 15 cm plate, respectively.

6. Spin cells immediately at 400 rcf for 3 min.

7. Carefully remove PBS and slowly but completely resuspend in 350 µl of ChIP lysis buffer containing protease inhibitors. At this step the chromatin can be frozen using liquid nitrogen and stored at −80°C for long-term storage. When using frozen samples, remove tubes from freezer and let thaw at 4°C before proceeding to the next step.

8. Shear DNA by sonicating the chromatin in order to produce fragments of an average length of 1 kb (*see* **Note 6**). This step is highly dependent upon the apparatus used for sonication, the cell type, and the amount of chromatin. The chromatin obtained from one 10 or 15 cm culture plate can generally be sonicated using a water-base sonicator (Bioruptor) at maximal intensity for 8–10 min (cycles of 30 s of sonication and 30 s of resting time).

9. Spin for 15 min at 15,800 rcf at 4°C. You should have a small but detectable pellet of insoluble cellular debris. Isolate the supernatant, which represents the total soluble chromatin. Samples can be frozen and stored at −80°C at this step using liquid nitrogen. Set aside a small aliquot (12 µl from 350 µl (~5%)) as input and leave in freezer. For sonication optimization, a small aliquot of the DNA (5–10 µl) can be incubated overnight at 65°C to reverse the formaldehyde cross-linking and purified using the Quiaquick PCR purification kit before being loaded on an agarose gel for fragment size verification. Meanwhile, the samples can be kept at 4°C or frozen.

10. Dilute the samples to 1:10 (at least 1:5) in ChIP dilution buffer.

11. Add the A/G Dynal magnetic beads pre-bound to the antibody of choice. Incubate overnight at 4°C under rotation.

12. Collect magnetic beads using a magnetic concentrator and wash six times with 500 µl of RIPA buffer. Incubate on mixer for 10 min between every second wash.

13. Wash twice with 500 µl of TE buffer (pH 7.6). The beads might not stick very well to the side of the tube at this point and it might be required to remove the TE buffer very carefully using a pipette.

14. Carefully remove all residual TE after the last wash.

15. Add 100 μl of ChIP elution buffer. Do the same to the input samples that were kept frozen. Vortex beads in this solution and incubate at 65°C overnight to reverse cross-linking.

16. Purify the DNA using the QIAquick PCR purification kit following the manufacturer's instructions. Briefly, add 500 μl of PB buffer, vortex, quickly spin down, and collect beads using a magnetic concentrator. Transfer the supernatant to the column and spin for 1 min at 15,800 rcf. Add 500 μl of PE buffer and spin twice for 1 min at 15,800 rcf. Elute using 30–50 μl of EB buffer for ChIPed DNA and 100 μl for input DNA.

17. Define enrichment of known target site(s) by real-time PCR (*see* **Note 7**). An aliquot of DNA is diluted from 1:2 to 1:10 and 5 μl is used in the following real-time PCR reaction in a total volume of 20 μl:

    5 μl DNA

    4 μl H_2O

    1 μl Primers (5 μM each)

    10 μl 2X SYBR

    The amount of ChIPed DNA is normalized to the input and enrichment is determined using a negative control, i.e., a region that is not bound by the factor of interest. Alternatively, the enrichment can be calculated as the relative concentration of a given chromatin fragment within an equivalent amount of ChIPed or input DNA. This first requires determining the concentration of DNA in those samples as described in the next step.

3.2. Preparation of the DNA for Hybridization onto the Tiling Arrays

1. Quantify ChIPed and input DNA using PicoGreen according to the manufacturer's recommendations using 5–10 μl of ChIPed DNA and 1 μl of input.

2. Take between 1 and 2 ng of ChIPed and input DNA, make up the volume to 200 μl with TE (pH 7.5), and add RNase A to a final concentration of 0.2 μg/μl. Incubate at 37°C for 1–2 h.

3. Add proteinase K to a final concentration of 0.2 μg/μl and incubate at 55°C for 2 h.

4. Extract once with an equal volume (200 μl) of phenol: chloroform:IAA. Vortex the samples for 20 s and add to phase lock gels (quick spin phase lock gels before adding the samples). Spin for 1 min at 15,800 rcf at room temperature and transfer the upper phase to new tubes.

5. Add 30 μg of glycogen and 1:10 volume (20 μl) of 3 M NaAc. The glycogen helps visualize the DNA pellet in subsequent precipitations.

6. Add two volumes of 100% ethanol (EtOH) and incubate for 30 min at −80°C.

7. Spin at maximum speed for 15 min at 4°C and wash with 150 µl of 70% EtOH.

8. Spin at maximum speed for 10 min and air-dry the pellet briefly.

9. Resuspend in 55 µl of distilled H_2O.

10. Prepare the end filling mixture:
    On ice, add:
    11 µl of 10X T4 DNA pol Buffer

    0.5 µl of BSA (10 mg/ml)

    1 µl of 10 mM dNTP mix

    0.2 µl of T4 DNA pol (3 U/µl)

    and up to 110 µl using dH_2O

11. Straight from ice, add to pre-cooled PCR machine and run the program: 12°C for 20 min.

12. Add 11.5 µl of 3 M NaAc and 10 µg glycogen.

13. Extract once with an equal volume of phenol:chloroform:IAA using phase lock gels as described in Step 4.

14. Add 260 µl of 100% EtOH and incubate for 30 min at −80°C.

15. Spin, wash with 70% EtOH, and resuspend dried pellet in 25 µl of H_2O as described in Steps 7–9.

16. Prepare the linker ligation mixture (*see* **Note 8**):
    All on ice (including thawing of reagents), add:
    7.8 µl of H_2O

    10 µl of 5X ligase buffer

    6.7 µl of annealed linkers (15 µM) – (*see* **Note 9**)

    0.5 µl of T4 DNA ligase

17. Incubate at 16°C overnight.

18. Add 6 µl of 3 M NaAc and 130 µl of 100% EtOH and incubate for 30 min at −80°C.

19. Spin, wash with 70% EtOH, and resuspend dried pellet in 25 µl of H_2O as described in Steps 7–9.

20. Prepare the PCR mixture:
    25 µl of DNA sample

    4.75 µl of H_2O

    4 µl of 10X Thermopol buffer

    5 µl of 2.5 mM dNTP mix

    1.25 µl of 40 µM long oligonucleotide

21. Place the tubes in a PCR machine and run the following program:
    55°C for 4 min

    72°C for 3 min

    95°C for 2 min

    (95°C for 30 s; 60°C for 30 s; 72°C for 1 min) × 25 cycles

    72°C for 4 min

22. Purify the DNA using the QIAquick PCR purification kit following the manufacturer's recommendations and elute in 60 μl of EB buffer.

23. Run 2 μl of each sample on a 2% agarose gel to check for amplification. A smear (generally between 400 and 200 bp) should be apparent. No prevalent bands should be detected apart from the smear.

24. Take a 2 μl aliquot and dilute it 1:10 for real-time PCR validation of enrichment(s). Repeat the RT-PCR performed before DNA amplification (last step of previous section) and make sure that the positive control regions are still enriched compared to negative control sites.

25. Prepare the fragmentation mixture (*see* **Note 10**):
    2 μl of diluted DNAseI – (*see* **Note 11**)

    5.5 μl of NEB buffer #4

    47.5 μl of amplified DNA (up to 6 μg)

26. Incubate the samples as follow:
    37°C for 30 min

    95°C for 15 min

27. Run 5 μl of each sample on a 2% agarose gel to check for good fragmentation. A smear should be apparent between 50 bp and 100 bp. If the fragmentation is not complete enough, 2 μl of diluted DNAseI can be added to the sample(s) and the digestion extended for a few minutes.

28. Prepare the labeling mixture (*see* **Note 12**):
    13 μl of TdT buffer

    1 μl of Biotin (1 mM)

    1 μl of TdT (30 U/μl)

    50 μl of fragmented DNA

29. Incubate the samples as follow:
    37°C for 16 h

    95°C for 10 min

30. Submit the ChIPed and input DNA for hybridization to a facility that owns the required equipment. Typically, 2 μg of amplified DNA are required per array.

31. Use MAT *(8)* and/or other algorithms to analyze the raw data (CEL files).

4. Notes

1. We have compared DNA amplification using LM-PCR and whole-genome amplification (using WGA kit from Sigma). While both were equally efficient regarding the identification of strongly enriched regions, LM-PCR allowed for the identification of a greater number of regions enriched at lower levels. In contrast, others have suggested that WGA produces less noise in the ChIP-chip signal compared to LM-PCR *(9)*.

2. Alternatively, Protein A/G Sepharose beads and not magnetic beads can be used even though they have a tendency to produce higher background (non-specific binding of chromatin to the beads).

3. A growing number of companies producing antibodies are offering ChIP-grade antibodies. If possible, comparing the efficiency of several antibodies directed against the protein of interest is generally recommended. When using antibodies that were never used in ChIP experiments, investigators can perform a western blot on the immunoprecipitate to make sure that the protein of interest is efficiently immunoprecipitated under the cross-linking conditions used. In addition, mass spectrometry may be performed to ensure the specificity of the signal.

4. Cross-linking can also be performed at room temperature. Differences in the ChIP efficiency of some factors have been reported depending on whether the cross-linking step is performed at room temperature or 37°C *(2)*. Even though a 10 min incubation with 1% formaldehyde is the most commonly used condition, these parameters can be modified to optimize a protocol. Similarly, for factors indirectly recruited to the DNA, the use of an additional protein–protein cross-linker (dual cross-linking) may be appropriate. This involves first applying a protein–protein cross-linker such as DSG for 30 min that is then washed away and replaced by a 1% formaldehyde solution for 10 min *(1, 2)*.

5. Some researchers have reported the use of organs rather that cells grown in vitro. Specific protocols to prepare cross-linked cells from living mouse organs have been published *(10, 11)*.

6. The DNA may also be fragmented through enzymatic reaction. When ChiPing for histones and histone modifications, investigators may digest the DNA with MNase in order to

generate ~140 bp fragments corresponding to the DNA wrapped around histones forming the nucleosome that is protected from digestion *(12)*.

7. Known positive and negative control regions may not necessarily be known. In this case, selective enrichment of bound chromatin sites cannot be checked and one must proceed to DNA amplification, labeling, and hybridization in order to know if regions were enriched in the ChIPed samples. These regions might then be used as controls in subsequent replication of the experiment.

8. A DNA phosphorylation step using T4 polynucleotide kinase can be performed before this step to improve linker ligation efficiency.

9. Annealing of linkers:

    Long oligonucleotide   -GCGGTGACCCGGGAGATCT GAATTC-

    Short oligonucleotide -GAATTCAGATC-

    Prepare the linker mixture: 250 µl Tris-HCl (1 M) pH 7.9

    375 µl of long oligonucleotide (40 µM stock)

    375 µl of short oligonucleotide (40 µM stock)

    Heat at 95°C for 5 min in water, remove from heat block, and let the linker mixture slowly cool on the bench. Prepare aliquots and store frozen.

10. The fragmentation is required here because this protocol described the preparation of DNA for hybridization onto Affymetrix tiling arrays that are characterized by short probes. Other platforms may not require further fragmentation of the DNA.

11. Dilute 1 µl of DNAseI (Affymetrix) in 31 µl of Tris 10 mM pH 7.8 to prepare the diluted DNAseI solution.

12. Using Affymetrix tiling arrays, both ChIPed and input samples are labeled using biotin and will be hybridized on different arrays. The signals given by the two samples are subsequently computationally compared. On the other hand, other platforms can use different labeling strategies allowing for simultaneous hybridization of ChIPed and input samples on the same array.

Acknowledgments

The authors are indebted to Drs. Jason Carroll and Timothy Geistlinger for advice on the ChIP-chip procedure.

References

1. Zeng PY, Vakoc CR, Chen ZC, Blobel GA, Berger SL (2006) In vivo dual cross-linking for identification of indirect DNA-associated proteins by chromatin immunoprecipitation. *Biotechniques* **41**:694, 696, 698

2. Nowak DE, Tian B, Brasier AR (2005) Two-step cross-linking method for identification of NF-kappaB gene network by chromatin immunoprecipitation. *Biotechniques* **39**:715–725

3. Johnson DS, Li W, Gordon DB, Bhattacharjee A, Curry B, Ghosh J, Brizuela L, Carroll JS, Brown M, Flicek P, Koch CM, Dunham I, Bieda M, Xu X, Farnham PJ, Kapranov P, Nix DA, Gingeras TR, Zhang X, Holster H, Jiang N, Green RD, Song JS, McCuine SA, Anton E, Nguyen L, Trinklein ND, Ye Z, Ching K, Hawkins D, Ren B, Scacheri PC, Rozowsky J, Karpikov A, Euskirchen G, Weissman S, Gerstein M, Snyder M, Yang A, Moqtaderi Z, Hirsch H, Shulha HP, Fu Y, Weng Z, Struhl K, Myers RM, Lieb JD, Liu XS (2008) Systematic evaluation of variability in ChIP-chip experiments using predefined DNA targets. *Genome Res* **18**:393–403

4. Lupien M, Eeckhoute J, Meyer CA, Wang Q, Zhang Y, Li W, Carroll JS, Liu XS, Brown M (2008) FoxA1 translates epigenetic signatures into enhancer-driven lineage-specific transcription. *Cell* **132**:958–970

5. Wang Q, Li W, Liu XS, Carroll JS, Janne OA, Keeton EK, Chinnaiyan AM, Pienta KJ, Brown M (2007) A hierarchical network of transcription factors governs androgen receptor-dependent prostate cancer growth. *Mol Cell* **27**:380–392

6. Carroll JS, Meyer CA, Song J, Li W, Geistlinger TR, Eeckhoute J, Brodsky AS, Keeton EK, Fertuck KC, Hall GF, Wang Q, Bekiranov S, Sementchenko V, Fox EA, Silver PA, Gingeras TR, Liu XS, Brown M (2006) Genome-wide analysis of estrogen receptor binding sites. *Nat Genet* **38**:1289–1297

7. Carroll JS, Liu XS, Brodsky AS, Li W, Meyer CA, Szary AJ, Eeckhoute J, Shao W, Hestermann EV, Geistlinger TR, Fox EA, Silver PA, Brown M (2005) Chromosome-wide mapping of estrogen receptor binding reveals long-range regulation requiring the forkhead protein FoxA1. *Cell* **122**:33–43

8. Johnson WE, Li W, Meyer CA, Gottardo R, Carroll JS, Brown M, Liu XS (2006) Model-based analysis of tiling-arrays for ChIP-chip. *Proc Natl Acad Sci USA* **103**:12457–12462

9. O'Geen H, Nicolet CM, Blahnik K, Green R, Farnham PJ (2006) Comparison of sample preparation methods for ChIP-chip assays. *Biotechniques* **41**:577–580

10. Chaya D, Zaret KS (2004) Sequential chromatin immunoprecipitation from animal tissues. *Methods Enzymol* **376**:361–372

11. Ray S, Das SK (2006) Chromatin immunoprecipitation assay detects ERalpha recruitment to gene specific promoters in uterus. *Biol Proced Online* **8**:69–76

12. Barski A, Cuddapah S, Cui K, Roh TY, Schones DE, Wang Z, Wei G, Chepelev I, Zhao K (2007) High-resolution profiling of histone methylations in the human genome. *Cell* **129**:823–837

Chapter 12

ChIP-Chip: Algorithms for Calling Binding Sites

X. Shirley Liu and Clifford A. Meyer

Abstract

Genome-wide ChIP-chip assays of protein–DNA interactions yield large volumes of data requiring effective statistical analysis to obtain reliable results. Successful analysis methods need to be tailored to platform specific characteristics such as probe density, genome coverage, and the nature of the controls. We describe the use of the respective software packages MAT and MA2C for the analysis of ChIP-chip data from one-color Affymetrix and two-color NimbleGen or Agilent tiling microarrays.

 Key words: ChIP-chip, probe modeling, normalization, peak detection.

1. Introduction

The ChIP-chip experiment is an effective way of investigating protein–DNA interactions and chromatin structure in vivo. The technique was first applied successfully to identify binding sites of transcription factors in budding yeast (1–3) and later to small portions of mammalian genomes (4). High-density DNA oligonucleotide tiling arrays that map to all nonrepetitive genomic regions have enabled more comprehensive views of mammalian chromatin biology (5). We focus on the analysis of ChIP-chip data on two of the most commonly used tiling array platforms for ChIP-chip, Affymetrix and NimbleGen, emphasizing methods that have been broadly used and have been shown to be effective. MAT (6) is a freely available and open source software package that has been developed for the analysis of Affymetrix tiling array data and successfully applied to many important studies (7–9). MA2C (10) provides a convenient and effective analysis tool for two-color NimbleGen or Agilent tiling array data.

Jonathan R. Pollack (ed.), *Microarray Analysis of the Physical Genome: Methods and Protocols, vol. 556*
© Humana Press, a part of Springer Science+Business Media, LLC 2009
DOI 10.1007/978-1-60327-192-9_12 Springerprotocols.com

2. Affymetrix Platform/MAT Software

2.1. Affymetrix Tiling Array Analysis

Affymetrix tiling arrays offer the highest probe density and the shortest (25 bp) oligonucleotide probe sequence. Short oligonucleotide probes tend to display large effects that are associated with probe sequence, which need to be modeled when measuring the DNA concentration of interest. Accounting for hybridization effects, handling outlier probe signals, and grouping similar probes for normalization are the key considerations in the analysis of this data type.

2.2. MAT: Model-Based Analysis of Tiling Arrays

MAT (6) offers an accurate, robust, flexible, and comprehensible analysis of Affymetrix tiling array data. MAT analysis consists of four elements: probe intensity standardization based on a probe sequence model; MATscore calculation using sliding windows of probes along the genome; p-value and false discovery rate (FDR) calculations; annotation of significant regions according to whether they are significant in the genome. MAT was designed to allow for the analysis of data from a variety of experimental setups: a single ChIP sample without replicates or controls, a single ChIP with a single control, and replicate ChIP samples with replicate controls. The analysis of a single ChIP is useful when developing a ChIP-chip protocol or selecting an antibody. It generates useful data and valuable information at a minimal cost. However, controls and replicates are important to reduce the number of false positives and may be required to generate high-quality data for publication.

2.3. Probe Intensity Standardization

In most ChIP-chip experiments the vast majority of probes register signal from genomic DNA that is not enriched in the ChIP experiment. To adjust for probe sequence-associated intensity effects, MAT uses a set of probes to estimate the parameters in a linear model using ordinary least squares regression. The model includes terms for each independent base at each position on the probe to account for the position-specific contribution of a base to the probe intensity. Probes with similar predicted intensities are grouped together in bins. As the variance in the intensity tends to increase with the intensity itself, MAT groups probes of similar sequence and standardizes each probe's intensity based on its model-predicted intensity and its bin variance. This standardization approach achieves probe background subtraction and normalization in one step. MAT is separately applied for each array and allows for the comparison of intensities between probes on different arrays and between different probes on the same array. Although MAT is based on the assumption that most probes measure only background signal, MAT produces good results even in cases where ChIP-enriched probes occupy a substantial percentage of the array (~20% in some histone modification ChIP-chip experiments).

2.4. Genomic Region Scoring

On the Affymetrix tiling array platform individual probes tend to produce somewhat unreliable information on the concentration of the hybridization DNA cocktail. A substantial number of probes may be unresponsive, having very low or very high intensities regardless of the experiment. Accuracy and robustness is achieved by combining information from multiple probes. MAT calculates a MATscore which is a function of the intensity and the number of probes within a range of genomic sequence, the extent of which is user specified. The default window size in MAT is 600 bp (2 × Bandwidth), a range which is compatible with the sonication fragment size for many ChIP-chip experiments. Although this window size is user specified, only in rare circumstances can an adjustment of this parameter be expected to have a significant impact on an experimental result. The precise definition of MATscore depends on whether a control array set is included in the analysis. If no control group is included MATscore is the trimmed mean of the probe intensities within that window divided by the square root of the number of probes within the window. When a control set of arrays is included it is the MATscore difference between treatment and control arrays. The trimmed mean is a robust statistic that excludes outlier probe signals and is defined as the mean after discarding a proportion of the highest and lowest signal values, which is usually set around 10%.

2.5. p-Value Estimation

MAT uses an empirical strategy to estimate the p-value for any window. First, the distribution of MATscores for a set of non-overlapping windows is examined. This distribution tends to have a longer tail on the right which represents the ChIP-enriched sites. The null distribution is assumed to be symmetrical about the mode and is estimated using the MATscores to the left of the mode. A p-value for a sliding window is calculated from that window's MATscore and the MATscore null distribution. Windows with p-values meeting a required significance level are merged if they overlap or are separated by less than a user-defined parameter MaxGap. MaxGap is often set to the same value as Bandwidth. Windows that contain few probes may be unreliable, so only windows containing more than the minimum number of probes (user parameter MinProbe, default value 10) are considered for peak calls.

2.6. FDR Estimation

MAT estimates a false discovery rate (FDR) that is the proportion of false positives within the set of significant regions. At each MATscore cutoff, there are often positive peaks above the cutoff, as well as "negative" peaks if the sign on the MATscores was to be reversed. When control samples are available, these "negative" peaks are simply regions where probes are higher in controls than in ChIP. As controls are not expected to give rise to biologically significant peaks, these "negative" peaks are considered false positives. The FDR is then simply the number of "negative" peaks divided by the number of positive peaks at each MATscore cutoff.

To define the set of regions that are to be called significant a user-specified cutoff needs to be set. This is done by specifying one of the parameters MATscore, p -value, or FDR. FDR is a good parameter to set as this is the most easily interpretable. When using a new antibody or protocol it is advisable to confirm the results of the ChIP-chip experiment using ChIP-PCR on sites with a range of *p*-values.

2.7. Downloading the Software

MAT is open source software and freely available for downloading at: http://chip.dfci.harvard.edu/~wli/MAT/

MAT is readily run on Linux-x86 and Mac OS X, and can run on other operating systems with some effort from the user. Instructions for installation are posted on the web site. Careful attention to the installation procedure and software requirements is necessary to ensure a smooth installation process. Sample data, including .cel files derived from a study of estrogen receptor binding on chromosomes 21 and 22 *(11)*, are available from the MAT website.

In addition to the software, two files are required for the ChIP-chip analysis:

1. .bpmap library files which contain the sequence, array coordinate to genome location map, and genome copy number of each probe. The .bpmap files that may be downloaded from the MAT website are different from those available from Affymetrix. The probe sequences in the MAT .bpmap file have been remapped to the reference genome build and the number of exact matches to the genome has been recorded. Redundant mappings of probes to within the same 1 kb region of the genome have been filtered out.

2. A repeat-library file which contains the chromosome coordinates of RepeatMasker repeats, simple repeats, and segmental duplication.

2.8. Running MAT

The following steps are recommended for data organization to run MAT.

1. Create a new directory, chip, and three subdirectories: chip/cel, chip/library, and chip/work. chip/cel and chip/work must be user readable and writeable while chip/library needs to be user readable but not necessarily writable. This is a mere recommendation, the precise names and organization of files is at the user's discretion.

2. Copy the raw data containing .cel files into the chip/cel directory. MAT requires all .cel files used in an analysis to be located in the same directory.

3. Copy the installation-provided sample .tag file to the `chip/work` directory, name the `my_chip.tag`, and edit the MAT parameters. Details about the `.tag` file content are described below.

4. Change directory to `chip/work` and run MAT from the command line

   `MAT my_chip.tag`

2.9. Tag File Definition

The .tag file is a plain text file that can be generated in any text editor such as emacs, vi, or notepad. The file is split into sections by the tokens "[data]", "[bpmap]", "[cel]", "[intensity analysis]", and "[interval analysis]". The "data" section specifies the names of the directories containing the .bpmap, .cel, and repeat library files. In this section the treatment and control groups are specified by a string of zeros and ones after the token "Group". Ones stand for ChIP .cel files, i.e., .cel files containing data from an Chromatin ImmunoPrecipitation replicate, while zeros stand for input .cel file, i.e., .cel files containing control data. In the example (**Fig. 12.1**), Group = 1100 means that the first two .cel files specified in any line of the "[cel]" section contain ChIP data while the last two contain genomic input data. The parameter "Pair" defines the way in which input data is used as a control. When input and IP arrays are strictly matched, setting "Pair = 1" results in the normalized value for each input probe being subtracted from the matching probe in the IP array. In most situations it is recommended that "Pair" be left blank. In the "[bpmap]" section, the name of each .bpmap file to be used for each .cel file is specified. In the example the Affymetrix human chr21/22 tiling chip set consists of three chips "A", "B", and "C". There is a separate .bpmap file specified for each chip in the set. The numbering system in the "[cel]" section needs to be consistent with that of the "[bpmap]" section. In the "[intensity analysis]" section the `BandWidth` parameter specifies the window size that is used to group probes that lie within a genomic region spanning *twice* the `BandWidth`. Regions that are separated from each other by a distance less than the `MaxGap` parameter are merged. `MinProbe` specifies the minimum number of probes requires for a region to be called enriched. The cutoffs defining significant regions are set in the "[interval analysis]" section. The user can set the cutoff by "Matscore", "*p*-value", or "FDR". Only one of these should be specified.

2.10. Output Files

MAT-generated files will appear in the `chip/work` directory. MAT returns two types of output file: the `.bar` files which contain the MATscore for each probe which can be imported into the Affymetrix Integrated Genome Browser, IGB, for visualization, and a `.bed` file with the chromosomal coordinates of all the

```
[data]
BpmapFolder = /home/jane/chip/library
CelFolder = /home/jane/chip/cel
GenomeGrp =
RepLib =
/home/jane/chip/library/Humanhg17Rep.lib
Group = 1100
Pair =

[bpmap]
1 = P1_CHIP_A.Anti-
Sense.hs.NCBIv35.NR.bpmap
2 = P1_CHIP_B.Anti-
Sense.hs.NCBIv35.NR.bpmap
3 = P1_CHIP_C.Anti-
Sense.hs.NCBIv35.NR.bpmap

[cel]
1 = MCF_ER_A1.CEL  MCF_ER_A2.CEL
MCF_INP_A1.CEL  MCF_INP_A2.CEL
2 = MCF_ER_B1.CEL  MCF_ER_B2.CEL
MCF_INP_B1.CEL  MCF_INP_B2.CEL
3 = MCF_ER_C1.CEL  MCF_ER_C2.CEL
MCF_INP_C1.CEL  MCF_INP_C2.CEL

[intensity analysis]
BandWidth = 300
MaxGap = 300
MinProbe  = 10
```

Fig. 12.1. Example of a MAT .tag file.

ChIP-regions with MATscore and repeat flags on the region labels which can be loaded into UCSC Genome Browser or IGB. Each row of the .bed file represents the chromosome start and end points, a label for a ChIP-enriched region, and a significance score, $-10 \log_{10}(p\text{-value})$. If more that 70% of a ChIP-enriched region is annotated as repetitive DNA through repeat masker, simple repeat *(12)* or segmental duplication *(13)* then "R_", "R_Si", or "R_SeN", respectively are appended to the region label. The appendix for segmental duplication "R_Se N" indicates that there are N duplications of the region in the genome. Repetitive DNA and segmental duplications are non-unique in the genome and often show up as false-positive peaks, and are therefore of limited value in subsequent analyses of the ChIP-chip data. An extended version of the .bed file, a .bed.xls file contains the following information for each ChIP-region: chromosome, start, end, name, $-10 \log_{10}(p\text{-value})$, MATscore, fold change, FDR(%), peak position, length. The peak position is identified as the probe with the highest MATscore. From the FDR_table.txt one can get a sense of the relationship between the FDR, the number of negative and positive peaks, the MATscore, and p-value.

3. NimbleGen Platform/MA2C Software

3.1. NimbleGen Tiling Array Analysis

The NimbleGen platform allows customers to design arrays that cover targeted genomic regions with any desired probe density that is compatible with the size of the array. The standard ChIP-chip protocol in the NimbleGen system involves hybridizing ChIP and control samples on the same array using a two-color system. Probes in the NimbleGen system are typically 35–70 bp in length and like the Affymetrix system display considerable variability in hybridization properties.

3.2. MA2C: Model-Based Analysis of Two-Color Arrays

MA2C (10) is a Java-based software package for the robust analysis of two-color tiling array data provided in the NimbleGen format. There are three key ways in which MA2C, in its treatment of two-color arrays, differs from MAT. First, probe behavior on the NimbleGen platform can be described adequately using probe GC content instead of the more sophisticated MAT model. Second, MA2C applies its probe sequence model to correct for the correlation between Cy3 and Cy5 channels in addition to adjusting for the GC-dependent mean and variance of each probe. Finally, as NimbleGen probe densities are lower and probe qualities higher than the Affymetrix platform, alternative methods are provided to score regions.

3.3. Probe Intensity Standardization

MA2C bins probes based on the number of G and C residues within an oligonucleotide probe. The MA2C model assumes that the paired background log Cy3 and Cy5 intensities follow a bivariate normal distribution that is dependent on the GC content. This model takes into account the correlation between Cy3 and Cy5 readings that tends to increase with increasing GC count. For each GC bin, MA2C computes the mean and variance for Cy3 and Cy5 probes independently, and the correlation between paired Cy3 and Cy5 measurements. MA2C has an option to calculate these model parameters using a generalization of Tukey's biweight estimation.

3.4. Genomic Region Scoring

The assignment of scores to regions, like MAT, is done using a series of windows of user-defined length, centered at each probe. A user-defined MA2Cscore may be defined as the median, pseudo-median, median polish, or trimmed mean of the probes in the window. The median and trimmed mean options are implemented by calculating the median or trimmed mean of all the probes in the window; when replicates are available, the median t-value or trimmed mean of all pooled probes in identical windows across replicates is used. The pseudo-median of a distribution is the median of all pairwise arithmetic means. Median polish is recommended for experiments with a large number of replicate samples, while trimmed mean is recommended for arrays with densely tiled probes. The pseudo-median and median provide robust alternatives that can be applied in experiments that are not densely tiled and have few available replicates.

3.5. FDR Estimation

The p-value and FDR estimation are carried out in MA2C using the same techniques as in MAT.

3.6. Downloading the Software

MA2C is open source software that runs on all platforms that support Java Runtime Environment 5.0 or higher and has been successfully tested on Linux-x86, Mac OS X, and Windows operating systems. MA2C can be downloaded for free from the following website together with installation instructions and user manual:

http://liulab.dfci.harvard.edu/MA2C/MA2C.htm

3.7. Using MA2C

The file structure of NimbleGen data consists of three main components, DesignFiles/, PairData/, and SampleKey.txt, which should all reside in the same parent directory. The text file SampleKey.txt, an example of which is provided in **Fig. 12.2**, contains the relevant design information about individual arrays; in particular, the file must contain DESIGN_ID, CHIP_ID, and DYE for each array. The directory DesignFiles/ contains the sequence (.ndf) and position (.pos,) files corresponding to each DESIGN_ID, while PairData/ contains the single channel data for each CHIP_ID.

To start MA2C on a Windows platform, double-click on MA2C\dist\MA2C.bat. To launch the program in Linux, Unix, or on a Mac: change directory to MA2C/dist/ and execute the command:

java -Xmx600m -jar MA2C.jar

To display the main interaction panel, click on the "run" button. Click on the "SampleKey" button to select a sample key file that either comes with a NimbleGen CD or is created by the user using a text editor. The entries in the sample key file are tab-delimited and must contain the CHIP ID, DESIGN ID, and DYE information corresponding to each experiment. **Figure 12.2** provides an example of a sample key file involving three chips for each of which the Cy5 channel is the IP and the Cy3 the genomic input control.

CHIP_ID	DYE	SAMPLE_DESCRIPTION	DESIGN_NAME	DESIGN_ID
49875	Cy3	genomic input	2005-04-25_HG17_50mer	1944
49875	Cy5	estrogen receptor IP	2005-04-25_HG17_50mer	1944
49880	Cy3	genomic input	2005-04-25_HG17_50mer	1944
49880	Cy5	estrogen receptor IP	2005-04-25_HG17_50mer	1944
49883	Cy3	genomic input	2005-04-25_HG17_50mer	1944
49883	Cy5	estrogen receptor IP	2005-04-25_HG17_50mer	1944

Fig. 12.2. Example of a sample key file used by MA2C to specify treatment and control channels.

If your data structure follows the pattern described above, MA2C will automatically look for sequence (.ndf, .pos) files and pair data (.txt) files and display the information on the GUI. If the sequence and pair data folder names do not follow the above convention, you can select the folders manually via the "Sequence" and "Pair Data" buttons. It is important that all data files in PairData/ have the .txt extension.

3.8. Normalization   In the SampleKey table, click Ctrl-left to select the ChIP DYE for each array to be normalized. Only the ChIP channel should be selected, as MA2C will automatically find and use the correct Input channel based on your selection of IP channel. In the example given in **Fig. 12.3**, to normalize all three data sets, one should select only the lines highlighted in grey. Choose a

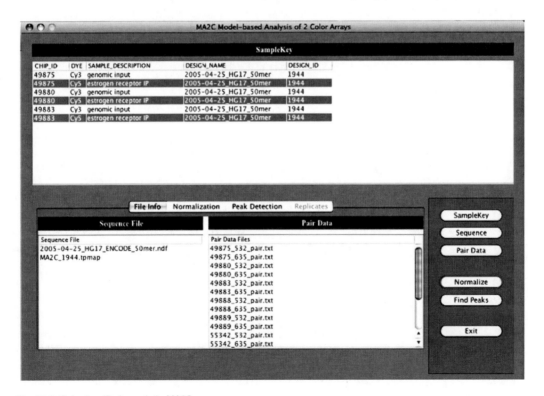

Fig. 12.3. Selecting IP channels in MA2C.

normalization method in the "Normalization" tab and click on the "Normalize" button to normalize the data. A `.tpmap` file will be generated in the sequence directory, and for each experiment, a `_raw.txt` and a `_normalized.txt` file will be created in the paired data directory. These files are used by MA2C but will not be of interest to the typical MA2C user. A new directory called `MA2C_Output` will be created in the same parent directory as `SampleKey.txt`.

3.9. Peak Detection

After normalization, check the "Peak Detection" tab for the available options that include the MA2C score statistic, the threshold statistic, the threshold value, and whether or not datasets are to be treated as replicates. Click the "Find Peaks" button to run the analysis. You do not need to renormalize the data each time you want to find peaks using different options. MA2C will generate the following files in the `MA2C_Output` directory: `MA2C_dataset.bed`, `MA2C_dataset.xls`, `MA2C_dataset.pdf`, `MA2C_dataset_FDRtable.txt`, `MA2C_dataset.MA2Cscore.bar`, and `MA2C_dataset.ratio.bar`, where `dataset` corresponds to a string of CHIP_IDs used to define the experiment. The `.bar` and `.xls` files contain the ChIP-enriched genomic regions. The `.pdf` files contain MA2C score, probe intensity ratios, and *p*-value histograms as well plots of the difference between the number of positive and negative peaks as a function of the FDR. The `.FDRtable.txt` allows the user to examine the relationship between the FDR, the number of negative and positive peaks, the MA2C score, and the *p*-value. The `.bar` files can be viewed in the Affymetrix Integrated Genome Browser. MA2C scores are contained in the `.MA2Cscore.bar` file and normalized probe ratios in the `.ratio.bar` file.

4. Conclusions

ChIP-chip data from tiling arrays can provide valuable biological insights if it is analyzed using robust statistical methods. MAT is a freely available, effective, software package for the analysis of Affymetrix tiling array data. MAT can be used to interpret data from a variety of experimental designs including a single ChIP with no replicates and no controls, a feature that is useful in the development of a ChIP-chip protocol and antibody selection. MA2C is a free, robust, analysis tool for NimbleGen two-color tiling array data that includes several diagnostic features for data quality assessment. Users need to join Google group to obtain a username and password for downloading, posting questions, and obtaining update emails.

References

1. Ren, B., Robert, F., Wyrick, J.J., Aparicio, O., Jennings, E.G., Simon, I., Zeitlinger, J., Schreiber, J., Nannett, N., Kanin, E., Volkert, T.L., Wilson, C.J., Bell, S.R., Young, R.A. (2000) Genome-wide location and function of DNA binding proteins. *Science* **290**, 2306–2309

2. Iyer, V.R., Horak, C.E., Scafe, C.S., Botstein, D., Snyder, M., Brown, P.O. (2001) Genomic binding sites of the yeast cell-cycle transcription factors SBF and MBF. *Nature* **409**, 533–538

3. Lieb, J.D., Liu, X., Botstein, D., Brown, P.O. (2001) Promoter-specific binding of Rap1 revealed by genome-wide maps of protein–DNA association. *Nat Genet.* **208**, 327–334

4. Horak, C.E., Mahajan, M.C., Luscombe, N.M., Gerstein, M., Weissman, S.M., Snyder, M. (2002) GATA-1 binding sites mapped in the β-globin locus by using mammalian ChIP-chip analysis. *Proc Natl Acad Sci.* **99**, 2924–2929

5. Cawley, S., Bekiranov, S., Ng, H.H., Kapranov, P., Sekinger, E.A., Kampa, D., Piccolboni, A., Sementchenko, V., Cheng, J., Williams, A.J., Wheeler, R., Wong, B., Drenkow, J., Yamanaka, M., Patel, S., Brubaker, S., Tammana, H., Helt, G., Struhl, K., Gingeras, T.R. (2004) Unbiased mapping of transcription factor binding sites along human chromosomes 21 and 22 points to widespread regulation of noncoding RNAs. *Cell* **116**, 499–509

6. Johnson, W., Li, W., Meyer, C. Gottardo, R., Carroll, J., Brown, M., Liu, X. (2006) Model-based analysis of tiling-arrays for ChIP-chip. *Proc Nat Acad Sci USA* **103**, 12457–12462

7. Carroll, J.S., Liu, X.S., Brodsky, A.S., Li, W., Meyer, C.A., Szary, A.J., Eeckhoute, J., Shao, W., Hestermann, E.V., Geistlinger, T.R., Fox, E.A., Silver, P.A., Brown, M. (2005) Chromosome-wide mapping of estrogen receptor binding reveals long-range regulation requiring the forkhead protein FoxA1. *Cell* **122**, 33–43

8. Zheng, Y., Josefowicz, S.Z., Kas, A., Chu, T.T., Gavin, M.A., Rudensky, A.Y. (2007) Genome-wide analysis of Foxp3 target genes in developing and mature regulatory T cells. *Nature* **445**, 936–940

9. Wendt, K.S., Yoshida, K., Itoh, T., Bando, M., Koch, B., Schirghuber, E., Tsutsumi, S., Nagae, G., Ishihara, K., Mishiro, T., Yahata, K., Imamoto, F., Aburatani, H., Nakao, M., Imamoto, N., Maeshima, K., Shirahige, K., Peters, J.M. (2008) Cohesin mediates transcriptional insulation by CCCTC-binding factor. *Nature* **451**, 796–801

10. Song, J.S., Johnson, W.E., Zhu, X., Zhang, X., Li, W., Manrai, A.K., Liu, J.S., Chen, R., Liu, X.S. (2007) Model-based analysis of two-color arrays (MA2C). *Genome Biol.* **8** Article R178, http://genomebiology.com/2007/8/8/R178

11. Carroll, J.S., Meyer, C.A., Song, J., Li, W., Geistlinger, T.R., Eeckhoute, J., Brodsky, A.S., Keeton, E.K., Fertuck, K.C., Hall, G.F., Wang, Q., Bekiranov, S., Sementchenko, V., Fox, E.A., Silver, P.A., Gingeras, T.R., Liu, X.S., Brown, M. (2006) Genome-wide analysis of estrogen receptor binding sites. *Nat Genet.* **38**, 1289–1297

12. Benson, G. (1999) Tandem repeats finder: a program to analyze DNA sequences. *Nucleic Acids Res.* **27**, 573–580.

13. Bailey, J.A., Yavor, A.M., Massa, H.F., Trask, B.J., Eichler, E.E. (2001) Segmental duplications: organization and impact within the current human genome project assembly. *Genome Res.* **11**, 1005–1017.

Chapter 13

Mapping Regulatory Elements by DNaseI Hypersensitivity Chip (DNase-Chip)

Yoichiro Shibata and Gregory E. Crawford

Abstract

Historically, the simplest method to robustly identify active gene regulatory elements has been enzymatic digestion of nuclear DNA by nucleases such as DNaseI. Regions of extreme chromatin accessibility to DNaseI, commonly known as DNaseI hypersensitive sites, have been repeatedly shown to be markers for all types of active *cis*-acting regulatory elements, including promoters, enhancers, silencers, insulators, and locus control regions. However, the original classical method, which for over 25 years relied on Southern blot, was limited to studying only small regions of the genome. Here we describe the detailed protocol for DNase-chip, a high-throughput method that allows for a targeted or genome-wide identification of *cis*-acting gene regulatory elements.

Key words: DNase-chip, DNaseI hypersensitivity, DNA regulatory elements.

1. Introduction

Genomic DNA sequence by itself provides relatively little information regarding identifying the location of gene regulatory elements. However, understanding how and where eukaryotic DNA is packaged into nucleosomes and higher-order structures of chromatin provides much more detailed information regarding the process of gene regulation *(1)*. While the majority of the genome is sequestered from regulatory proteins, there are regions within chromatin where nucleosomes are displaced by transcription factors within noncoding functional elements. For almost three decades, DNaseI hypersensitivity (HS) assays have been used to successfully identify the location of these open chromatin regions within the genome. These regions have been shown to be associated with all types of regulatory elements, including promoters, enhancers,

Jonathan R. Pollack (ed.), *Microarray Analysis of the Physical Genome: Methods and Protocols, vol. 556*
© Humana Press, a part of Springer Science+Business Media, LLC 2009
DOI 10.1007/978-1-60327-192-9_13 Springerprotocols.com

silencers, insulators, and locus control regions *(2–6)*. The selectivity of DNaseI to enzymatically cleave at these regulatory regions is thought to be an order of magnitude higher than in regions of transcriptionally active genes and two orders of magnitude higher than other regions of bulk chromatin *(4)*.

Before the availability of whole-genome sequences as well as high-throughput genomic tools, individual DNaseI HS sites within small regions of the genome (10–20 kb) were identified using Southern blot assays *(7)*. However, this labor-intensive method was not readily scalable to study large chromosomal regions and entire genomes. With the availability of high-resolution tiled microarrays, DNase-chip method was developed to identify DNaseI HS sites within any region of interest, including the entire genome *(8–10)*.

Even as the availability of next-generation high-throughput sequencing technologies continue to become more widespread, the DNase-chip method continues to be a valuable tool to both complement and validate sequence data gathered from DNase-seq studies *(10)*. We have found that both DNase-chip and DNase-seq have similarly high sensitivity and specificity to identifying valid DNaseI HS sites *(10)*. An added advantage of DNase-chip is that it can be used to directly interrogate smaller targeted regions of the genome.

2. Materials

2.1. Cell Collection and Nuclei Prep

1. Phosphate-buffered saline (pH 8.0)
2. Puregene Core Kit A (Gentra/Qiagen)
3. Trizol reagent (Invitrogen)
4. RSB Buffer: 10 mM Tris-HCl pH 8.0, 10 mM NaCl, 3 mM MgCl$_2$
5. NP40 (Igepal CA-630, Sigma). Prepare a 10% solution.
6. Wide-bore pipette tips (or use clean razor blade to clip off pipette tip)
7. Trypan blue (Gibco)

2.2. DNaseI Digestion

1. 1% InCert low melt agarose (Lonza) in sterile 50 mM EDTA pH 8.0. Aliquot into 1.5 ml tubes and store at 4°C.
2. 50 mM EDTA pH 8.0
3. LIDS buffer: 1% (w/v) Lauryl sulfate lithium salt (Sigma), 10 mM Tris-HCl pH 8.0, 100 mM EDTA pH 8.0.
4. DNaseI (Roche)
5. Plug molds (Bio-Rad)
6. Screened Plug Caps to fit 50 ml conical tubes (Bio-Rad)

2.3. Pulsed-Field Gel Electrophoresis (CHEF)

1. Yeast Chromosome PFG molecular weight marker (NEB)
2. 0.5X TBE buffer
3. Agarose (Invitrogen)

2.4.–2.5. Blunt-End Randomly Sheared DNA and DNased DNA

1. DNA Polymerase Buffer: 50 mM NaCl, 10 mM Tris-HCl pH 8.0, 10 mM MgCl$_2$, 1 mM dithiothreitol (DTT). Prepare 1 l.
2. T4 DNA Polymerase (NEB)
3. Glycogen (Roche)
4. Ethyl alcohol (100 and 70%)
5. Phenol
6. Phenol:Chloroform:Isoamyl Alcohol
7. Chloroform

2.6. Ligation of Biotinylated Linkers

1. T4 Ligase (NEB)
2. Annealed linker set A: Oligos #1/#2 (Integrated DNA technologies, HPLC purified)
3. Oligo #1: 5′/5Bio/GCG GTG ACC CGG GAG ATC TGA ATT C -3′
4. Oligo #2: 5′/5Phos/GAA TTC AGA TC/3AmM/-3′
5. 5X Ligase Buffer (Invitrogen)

2.7. Shear DNA

1. Branson Sonicator
2. TE Buffer: 10 mM Tris-HCl pH 8.0, 1 mM EDTA pH 8.0

2.8. Bind to Streptavidin Beads

1. Dynal Streptavidin beads (Invitrogen Dynal M-280)
2. Dynal magnet (MPC-S)
3. Binding buffer: 10 mM Tris-HCl pH 8.0, 1 mM EDTA pH 8.0, 1 M NaCl

2.9. Blunt-End Sheared DNA Ends

1. T4 DNA Polymerase (NEB)
2. dNTP, 10 mM (Roche)

2.10. Ligation of Non-biotinylated Linkers to Sheared DNA Ends

1. T4 Ligase (NEB)
2. Annealed linker set B: Oligos #3/#4 (Integrated DNA technologies, HPLC purified)
3. Oligo #3: 5′ GCG GTG ACC CGG GAG ATC TGA ATT C -3′
4. Oligo #4: 5′ GAA TTC AGA TC -3′
5. 5X Ligase Buffer (Invitrogen)

2.11. Ligation-Mediated PCR (LM-PCR)

1. Taq DNA Polymerase (Invitrogen)
2. dNTP, 10 mM (Roche)

3. 10X ThermoPol Buffer (NEB)

4. Oligo #3

2.12. Purify LM-PCR Product

1. ArrayIT Microarray Probe Purification Kit (#FPP)

3. Methods

The DNase-chip method is outlined in **Fig. 13.1**. Before starting the protocol, cool centrifuge and buffers to 4°C. Melt 1% InCert Low Melt agarose (in 50 mM EDTA) at 75°C and keep melted at 55°C using heat block. Set water bath to 37°C. Make DNaseI dilutions in cold RSB, mix thoroughly, and aliquot into appropriate number of 1.5 ml tubes to be used for the DNaseI digestion step (keep on ice) (*see* **Note 1**).

3.1. Cell Collection and Nuclei Prep

1. Harvest 5×10^7 cells and centrifuge at 160g for 5 min at 4°C; carefully vacuum or pipette off the supernatant (do not pour off).

2. Separate 5–10×10^6 cells from the total harvest, and isolate genomic DNA (Gentra/Qiagen Puregene Core Kit A). This will be used for the randomly sheared genomic DNA control.

3. Wash remaining cells two times with 50 ml cold PBS. Resuspend final cell pellet in 500 μl cold RSB by gentle flicking and transfer to a 15 ml conical tube on ice. Wash the 50 ml tube with 500 μl cold RSB and combine with the first 500 μl cell suspension.

4. Slowly pour cold RSB + 0.1% NP40 into the 15 ml conical containing the cell suspension (total of 14 ml) to gently lyse the cells (*see* **Note 2**). Invert tube 5–10 times (do not pipet).

5. Spin down immediately at 500g for 10 min at 4°C to pellet nuclei. Completely remove supernatant using vacuum.

6. Resuspend nuclei pellet in 900 μl cold RSB buffer and mix (by flicking). Pellet should appear white and fluffy and should resuspend completely. Check a small sample of cells to confirm >99% of cells are Trypan blue positive (*see* **Note 3**).

3.2. DNaseI Digestion

1. Make DNaseI dilutions in RSB buffer and mix thoroughly. Add 12 μl of DNaseI dilutions to 1.5 ml eppendorf tubes and keep on ice. Always make fresh DNaseI dilutions (*see* **Note 4**).

2. Typical DNaseI working concentrations:

Tube#	Amount DNaseI	Dilution (in RSB)	Add to digestion reaction
1	0	0	0 (keep at 4°C – on ice)
2	0	0	0 (37°C)
3	0.12 U	1/1,000	12 μl
4	0.4 U	1/300	12 μl
5	1.2 U	1/100	12 μl
6	4.0 U	1/30	12 μl
7	12.0 U	1/10	12 μl

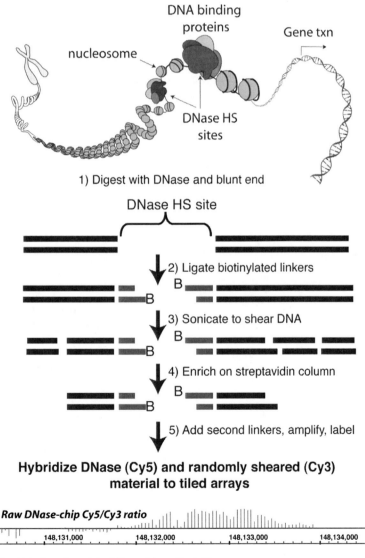

Fig. 13.1. Overview of the DNase-chip method *(9)*.

3. Pipette 120 μl of resuspended nuclei into the seven eppendorf tubes that contain the various DNaseI concentrations. Any pipeting from this point to the end of the protocol must use wide-bore pipette tips (or tips clipped with a clean razor blade) to minimize random shearing (*see* **Note 5**). Mix nuclei with DNaseI concentrations by gentle, but thorough, flicking (do not pipet).

4. Incubate in a 37°C water bath for 10 min (keep tube #1 on ice to monitor endogenous nuclease activity).

5. Add 330 μl of 50 mM EDTA to stop the reaction – total volume is now approximately 450 μl. Invert five times to mix (do not pipet), keep at room temperature, and immediately embed DNA.

6. Confirm the InCert low melt gel agarose is completely melted and at 55°C. Equilibrate the DNaseI-treated DNA to 55°C for 1 min.

7. Using a wide-bore pipette tip (or tip cut with razor), pipette 450 μl of InCert low melt gel agarose to nuclei/DNaseI tubes and invert four times to mix (do not pipette). Using same wide-bore pipette tip slowly pipette ~80 μl of DNaseI-treated nuclei/agarose mixture into Bio-Rad plug molds. Let it set at 4°C for 5 min to solidify.

8. Release plugs into 50 ml conical tubes containing 50 ml of LIDS buffer. Cover tubes with CHEF Screened Plug Caps before placing tube caps. Incubate 1–2 h at room temp while gently shaking at ~60 rpm (keep tubes on their side).

9. Incubate with fresh LIDS buffer overnight at 37°C (not shaking, but lay tubes on their side).

10. Wash plugs in 50 mM EDTA pH 8.0, five times 50 ml for 1 h each (room temp, shaking at ~60 rpm). Wash the plug caps and tube caps with distilled water at the fourth and fifth washings to remove excess detergent. No bubbles caused by residual detergent should be detected after fifth wash.

11. Store indefinitely at 4°C in 50 mM EDTA.

3.3. Pulsed-Field Gel Electrophoresis (CHEF)

1. These instructions assume the use of a Bio-Rad CHEF gel system, but any pulsed-field gel apparatus may be used. It is important to run the DNaseI-treated samples on a pulsed-field gel to determine the optimal levels of DNaseI digestion.

2. Prepare 3 l of 0.5X TBE and circulate and chill to 16°C in the CHEF system.

3. Prepare a 1% Agarose gel in 0.5X TBE for PFG electrophoresis.

4. Dry load one-third of the plugs into the wells of each lane. Use two small metal spatulas to help slide the plugs into the wells. Also load NEB Yeast Chromosome PFG marker to determine relative size of the fragments.

5. Pulse-field conditions: 20–60 s switch time for 18 h, 6 V/cm, Pump setting = 60 (~0.65L/min), Buffer temperature = 16°C.

6. There should be a gradual and consistent change in fragment sizes (smears) reflecting the increase in concentrations of DNaseI. Plugs from optimally digested samples will be used in the next steps. Examples of optimally digested samples are shown in **Fig. 13.2A**.

3.4. Blunt-End Randomly Sheared DNA

1. Randomly sheared genomic DNA is a necessary control required for microarray analysis to determine the baseline level of hybridization of non-DNase-enriched DNA fragments.

2. Isolate genomic DNA following the protocol from the Gentra/Qiagen kit.

3. Add water to bring total volume to 600 μl.

4. Extract DNA using 600 μl of Phenol/Chloroform/Isoamyl alcohol and vortex on high for 5 min to shear the DNA. Collect the aqueous layer.

5. Add 600 μl chloroform to the aqueous layer and vortex again for 5 min at high speed to further shear the DNA.

6. Ethanol precipitate DNA. Add one-tenth volume 3 M NaOAc, two volumes 100% EtOH, and 1 μl glycogen and place in −20°C for 30 min. Spin for 15 min at > 15,000g at 4°C. Wash DNA pellet with 1 ml 70% EtOH and spin again for 5 min.

7. Remove supernatant, quick spin, and remove all traces of liquid. Let dry for no more than 4 min to ensure DNA will go back into solution. Resuspend pellet in 40 μl of 10 mM Tris-HCl pH 8.0.

8. Blunt-end 40 μg of randomly sheared DNA for 1 h at room temperature, following the recipe:

10X Polymerase Buffer	20 μl
dNTPs (10 mM)	5 μl
BSA (10 mg/ml, 100X)	2 μl
T4 DNA Polymerase	1 μl
Randomly sheared DNA	(40 μg)
H$_2$O	(to 200 μl)

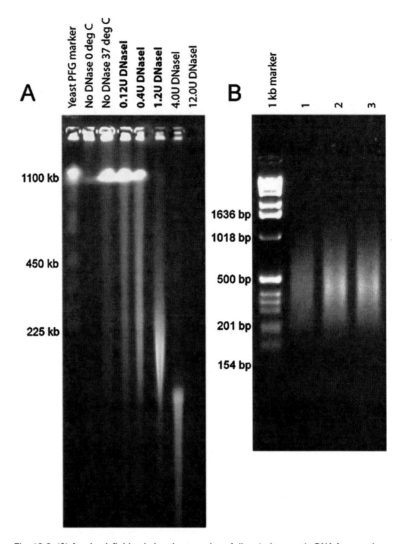

Fig. 13.2. (**A**) A pulsed-field gel showing a series of digested genomic DNA from various DNaseI concentration digestions obtained from HeLa cells. Note that DNaseI concentrations of 0.12, 0.4, and 1.2 U produced optimal smearing patterns and will be processed further to make the DNase-chip microarray hybridizing library. (**B**) A 2% agarose gel showing LM-PCR products that range between 200 and 700 bases. Lanes 1–3 were loaded with 10 μl from independent LM-PCR reactions.

9. Phenol/Chloroform/Isoamyl:Chloroform extract (using wide-bore p1000 tips trimmed with razor). Ethanol precipitate DNA and resuspend in 25 μl 10 mM Tris-HCl pH 8.0.

3.5. Blunt-End DNased DNA In-Gel

1. Wash optimal DNaseI-digested DNA plugs in 3 × 50 ml (1 h each wash) T4 DNA Polymerase Buffer to remove EDTA.

2. Use screened plug caps and vacuum to carefully remove all traces of liquid.

3. Blunt end the in-gel DNaseI-treated DNA for 3–4 h (gently mix every hour) at room temp, following the recipe:

DNA plug (low melt gel)	~80 μl (volume of plug)
10X Polymerase Buffer	12 μl
dNTPs (10 mM)	5 μl
T4 DNA Polymerase	6 μl
BSA (10 mg/ml, 100 ×)	2 μl
H₂O	100 μl

4. Transfer plugs to 1.5 ml eppendorf tubes and adjust volume to 600 μl with TE Buffer.

5. Heat to 65°C for 10 min. Monitor and flick tubes hard every couple of minutes to help dissolve agarose.

6. Phenol: Phenol/Chloroform/Isoamyl:Chloroform extract DNA (use wide-bore pipette tips).

7. Ethanol precipitate DNA, and air-dry pellet no longer than 4 min to ensure high-molecular-weight DNA will go back into solution. Resuspend in 25 μl 10 mM Tris-HCl pH 8.0.

3.6. Ligation of Biotinylated Linkers

1. Anneal oligos to create linkers. Mix 10 μl 1 M Tris-HCl pH 8.0, 10 μl 5 M NaCl, 2 μl 0.5 M EDTA, 375 μl of 40 μM oligo #1, 375 μl of 40 μM oligo #2, and 238 μl sterile water. Put mix in beaker full of boiling water for 2 min, remove beaker from heat, and let oligos slow-cool in water bath until room temperature is reached. Cool overnight at 4°C and aliquot/store at −20°C. *To prevent primers from becoming un-annealed, always thaw annealed linkers on ice.*

2. Ligate biotinylated linker set A to blunt-ended DNA (both DNaseI-treated and randomly sheared DNA):

DNA	(5 μg)
5X Ligase Buffer	10 μl (viscous: use wide-bore tips)
Annealed linker set A	6.7 μl (thaw linker on ice)
T4 Ligase	0.5 μl
H₂O	(up to 50 μl)

3. Incubate overnight at 16°C.

3.7. Shear DNA

1. These instructions assume the use of a Branson Sonicator using a setting of 3.

2. Transfer ligated DNA/linker to a 15 ml conical tube and bring total volume to 1.5 ml with TE Buffer and place on ice.

3. With the tubes placed in an ice bath, sonicate each sample 8 × 25 s. Sonicator tip should almost touch the bottom of the tube and should be cooled after each round in an ice bath.

3.8. Bind to Streptavidin Beads

1. Use 100 μl of Dynal Streptavidin bead suspension per reaction.

2. In 1.5 ml eppendorf tube wash beads 3 × 1 ml in Binding Buffer (TE + 1 M NaCl) using the Dynal magnet to capture the beads after each wash. Resuspend after last wash in appropriate amount of Binding Buffer.

3. Add 300 μl 5 M NaCl to sonicated DNA to make a final NaCl concentration of 1 M and transfer to a 1.5 ml eppendorf tube.

4. Add 100 μl of Dynal beads to sonicated DNA and rock at room temperature for 15 min.

5. Wash 3 × 1 ml in Binding Buffer, and finish with 1 × 1 ml wash in TE. Use pipette to carefully remove wash buffer after each wash step/bead capture – *beads will be very loose on magnet.*

6. After final wash, carefully remove traces of liquid with vacuum manifold.

3.9. Blunt-End Sheared DNA Ends

1. Add to Beads:

H_2O	97.3 μl
10X Polymerase Buffer	11 μl
dNTP (10 mM)	1 μl
BSA (100X)	1 μl
T4 DNA Polymerase	0.2 μl
Total:	~110 μl

2. Incubate for 1 h at 16°C and resuspend beads once during incubation.

3. Wash 3 × 1 ml in Binding Buffer, and 1 × 1 ml in TE.

4. After last wash, carefully remove traces of liquid with vacuum manifold.

3.10. Ligation of Non-biotinylated Linkers to Sheared DNA Ends

1. Add to Beads:

H$_2$O	32.8 µl
5X Ligase Buffer	10 µl (viscous: use wide-bore tips)
Annealed linker set B	6.7 µl (thaw linker on ice)
T4 Ligase	0.5 µl
Total:	50 µl

2. Resuspend beads and incubate overnight at 16°C.
3. Wash 3 × 1 ml in Binding Buffer, and 1 × 1 ml in TE.
4. Resuspend beads in 50 µl 10 mM Tris-HCl pH 8.0.

3.11. Ligation-Mediated PCR (LM-PCR)

1. Depending on the amount of desired PCR reaction product, prepare the appropriate number of PCR reactions.
2. Prepare PCR reaction:

H$_2$O	41 µl
10X ThermoPol Buffer	5 µl
dNTP (10 mM)	1.25 µl
Oligo #3 (40 µM)	1.25 µl
Beads	1 µl
Taq DNA Polymerase	0.5 µl
Total:	50 µl

3. Cycling Conditions:

95°C × 2 min
95°C × 30 s
60°C × 30 s } 25 cycles
72°C × 1 min
72°C × 5 min
4°C × ∞

4. Run 5–10 µl of the PCR product on a 2% agarose gel to confirm product sizes range between 200 and 700 bases (**Fig. 13.2B**).

3.12. Purify LM-PCR Product

1. Add TE Buffer to PCR product to bring volume up to 500 μl.
2. Phenol/Chloroform/Isoamyl:Chloroform extract DNA.
3. Ethanol precipitate DNA and resuspend pellet in 25 μl of TE Buffer.
4. Clean up DNA on ArrayIt columns.
5. Elute final product in 50 μl sterile water.
6. Combine LM-PCR products from multiple DNase concentrations (*see* **Note 4**).

3.13. Label and Hybridize to Arrays

1. Use standard dual-color ChIP-chip labeling (Cy5 for DNase-treated material and Cy3 for randomly sheared material) and hybridization protocols. We often send purified LM-PCR product to NimbleGen (Madison, WI) for labeling, hybridization, and scanning of tiled arrays (full service ChIP-chip).
2. Regions enriched for DNase signals can be identified using tiled array peak calling algorithms such as ACME or ChIPO-Tle *(11, 12)*.

4. Notes

1. The varying degrees of endogenous nuclease activity in different cell and tissue types may complicate the quality of DNase-chip data. Therefore it is important to keep samples and buffers on ice.

2. Optimization of NP-40 concentration: Care must be taken to avoid overlysing the cells while preparing cell nuclei prior to DNaseI digestion. NP-40 is a detergent that disrupts the cytoplasmic membrane but does not easily break the nuclear membrane. The nuclei pellet should appear white and should completely resuspend in RSB Buffer with modest agitation. When cells are overlysed, the lysate appears viscous and will not resuspend evenly. If this happens, the NP-40 concentration should be reduced to 0.05% or 0.01%. Trypan blue staining can be used to determine the minimal amount of NP40 necessary to achieve >99% lysis.

3. This protocol has been optimized for using 50 million cells. However, this number may be difficult to obtain for rare or difficult-to-grow cell types. We have found that as low as 5 million cells produce data comparable to studies using 50 million cells. If using a smaller number of starting cells (<50 million), several variables should be adjusted. First, consider reducing the number of plugs for the "No DNase"

controls from 10 plugs down to 1 plug since these controls are usually only used to monitor endogenous nuclease activity. Second, adding more diluted DNaseI concentrations and eliminating the highest concentrations of DNaseI from the digestion series will further reduce the required amount of nuclei. Third, one may reduce the number of plugs made for each DNase concentration. Reflecting these adjustments, the volume of nuclei resuspension buffer (RSB) in Step 6 in Section **3.1** can be reduced.

4. DNaseI HS sites are not binary, but instead represent a continuum of openness. Therefore, to capture both weak and strong DNaseI HS sites, we often pool material from 2 to 3 different DNaseI concentrations. It is often difficult to predict the optimal DNaseI concentrations, due to variations in cell heterochromatin, cell density, enzyme activity, incubation time, and inconsistent pipetting. Optimally digested samples can only be distinguished after running out a sample from each of the different DNaseI concentrations on PFG gel (**Fig. 13.2A**). Using material that is over-digested or under-digested results in low signal-to-noise ratios.

5. Care must be taken to avoid mechanical shearing of the samples anytime before the embedded DNA is blunt ended. Unwanted shearing of chromatin can introduce random breaks within chromatin and falsely introduce sites in the genome that may appear to be DNaseI HS sites. As noted in the protocol, only pipette when necessary, and always use wide-bore pipette tips. At least two biological replicates are suggested to confirm DNaseI HS sites are real.

Acknowledgments

We would like to thank Lingyun Song for her technical assistance and helpful discussions.

References

1. The Encode Project Consortium. (2007) Identification and analysis of functional elements in 1% of the human genome by the ENCODE pilot project. *Nature* **447**, 799–816.

2. Stalder, J., Larsen, A., Engel, J. D., Dolan, M., Groudine, M., and Weintraub, H. (1980) Tissue-specific DNA cleavages in the globin chromatin domain introduced by DNAase I. *Cell* **20**, 451–460.

3. Felsenfeld, G., and Groudine, M. (2003) Controlling the double helix. *Nature* **421**, 448–453.

4. Gross, D. S., and Garrard, W. T. (1988) Nuclease hypersensitive sites in chromatin. *Annu. Rev. Biochem.* **57**, 159–197.

5. Keene, M. A., Corces, V., Lowenhaupt, K., and Elgin, S. C. (1981) DNase I hypersensitive sites in Drosophila chromatin occur at the 5' ends of regions of transcription. *Proc. Natl. Acad. Sci. USA* **78,** 143–146.

6. McGhee, J. D., Wood, W. I., Dolan, M., Engel, J. D., and Felsenfeld, G. (1981) A 200 base pair region at the 5' end of the chicken adult [beta]-globin gene is accessible to nuclease digestion. *Cell* **27,** 45–55.

7. Wu, C. (1980) The 5' ends of Drosophila heat shock genes in chromatin are hypersensitive to DNase I. *Nature* **286,** 854–860.

8. Crawford, G. E. (2004) Identifying gene regulatory elements by genome-wide recovery of DNase hypersensitive sites. *Proc. Natl. Acad. Sci. USA* **101,** 992–997.

9. Crawford, G. E., Davis, S., Scacheri, P. C., Renaud, G., Halawi, M. J., Erdos, M. R., Green, R., Meltzer, P. S., Wolfsberg, T. G., and Collins, F. S. (2006) DNase-chip: a high-resolution method to identify DNase I hypersensitive sites using tiled microarrays. *Nat. Meth.* **3,** 503–509.

10. Boyle, A. P., Davis, S., Shulha, H. P., Meltzer, P., Margulies, E. H., Weng, Z., Furey, T. S., and Crawford, G. E. (2008) High-resolution mapping and characterization of open chromatin across the genome. *Cell* **132,** 311–322.

11. Scacheri, P. C., Crawford, G. E., and Davis, S. (2006) *in* Methods in Enzymology, Vol. 411, pp. 270–282, Academic Press, San Diego.

12. Buck, M., Nobel, A., and Lieb, J. (2005) ChIPOTle: a user-friendly tool for the analysis of ChIP-chip data. *Genome Biol.* **6,** R97.

Chapter 14

Microarray Analysis of DNA Replication Timing

Neerja Karnani, Christopher M. Taylor, and Anindya Dutta

Abstract

Although all of the DNA in an eukaryotic cell replicates during the S-phase of cell cycle, there is a significant difference in the actual time in S-phase when a given chromosomal segment replicates. Methods are described here for generation of high-resolution temporal maps of DNA replication in synchronized human cells. This method does not require amplification of DNA before microarray hybridization and so avoids errors introduced during PCR. A major advantage of using this procedure is that it facilitates finer dissection of replication time in S-phase. Also, it helps delineate chromosomal regions that undergo biallelic or asynchronous replication, which otherwise are difficult to detect at a genome-wide scale by existing methods. The continuous TR50 (time of completion of 50% replication) maps of replication across chromosomal segments identify regions that undergo acute transitions in replication timing. These transition zones can play a significant role in identifying insulators that separate chromosomal domains with different chromatin modifications.

Key words: Replication timing, BrdU labeling, tiling arrays, TR50.

1. Introduction

DNA replication is a key event in the cell cycle, occurring within a confined period termed S-phase *(1)*. Analysis of replication time for individual genes or chromosomal regions has historically relied on either fractionation of S-phase followed by semi-quantitative PCR for synthesis-induced increase in copy number *(2, 3)* or the counting of FISH signals in S-phase nuclei *(4)*. Although, these methods show local variation in replication timing, the laborious nature of the methods have restricted high-resolution analysis to small regions of the chromosomes *(3)* or lower-resolution analysis to single chromosomal segments *(2)*. The completion of the human genome sequence and the advent of genome-tiling microarrays have

Jonathan R. Pollack (ed.), *Microarray Analysis of the Physical Genome: Methods and Protocols, vol. 556*
© Humana Press, a part of Springer Science+Business Media, LLC 2009
DOI 10.1007/978-1-60327-192-9_14 Springerprotocols.com

provided an opportunity to study time of replication at a much finer resolution. Here, we detail methods for studying temporal behavior of replication at 25 bp resolution by using nucleotide analog incorporation, density centrifugation, and hybridization. This approach relies on the synchronization of cells to obtain the replication pattern from multiple discrete intervals of S-phase. This strategy has advantages over the existing S:G1 ratio based method of mapping replication timing. In a typical S:G1 method, DNA content is examined at the end and beginning of S-phase and the ratio between these measurements is used to estimate the time of replication. Thus an early replicating segment has a ratio that is closer to 2, while a late replicating segment has a ratio that is closer to 1 *(5)*. In such an experiment, segments showing biallelic replication appear to replicate near the middle of S-phase and provide misleading results. In the method described here, computing signal enrichment for multiple time intervals in S-phase allows a finer dissection of the temporal profile of replication and also identifies biallelic or asynchronously replicating regions of the genome more accurately *(6)*.

Most genomic methods depend on amplification of the experimental material prior to hybridization; this can change relative amounts of DNA in a complex mixture, and can hence provide unreliable signal values. In the strategy described here, there is no amplification step prior to hybridization, thus freeing the method from any artifacts that could arise due to amplification bias.

We also provide a section in this chapter on the algorithms used to generate continuous TR50 curves along the length of the chromosome. TR50 plots can be used to segregate discrete regions of early, mid, late, and Pan-S replication. Additionally, these curves provide information on genomic regions that undergo acute transition in replication time. These replication domains can be valuable indicators of chromatin structure, as we have observed them to have different levels of gene expression as well as activating and repressing histone marks *(6)*. Finally, the local minima of the TR50 curve show areas that replicate earlier than the flanking regions and so are likely to contain origins of replication, as has been shown previously in *Saccharomyces cerevisiae, (7)*. Thus, the hundreds of minima in the TR50 profile are likely to be at or near origins of replication.

2. Materials

2.1. Cell Culture, Synchronization, BrdU Labeling, and FACS

1. Cell culture: Dulbecco's Modified Eagle's Medium (DMEM; CELLGRO) supplemented with 10% iron supplemented donor calf serum (CELLECT) and 1% Penicillin-Streptomycin (GIBCO).

2. Synchronization: 1 M Thymidine (Sigma) prepared in phosphate-buffered saline (PBS). 1 μg/μl Aphidicolin (Sigma) dissolved in DMSO.

3. Bromodeoxyuridine (BrdU; Sigma) is dissolved in PBS at 10 mM concentration. Filter sterilized using 0.2-μm filter (CORNING) and stored in aliquots in dark at −20°C.

4. Propidium iodide FACS: Dilute 1 mg propidium iodide (Sigma) in 10 ml sterile water, 25 μl of 20% NP40, and 10 μl of 10 mg/ml RNase A (Roche Applied Sciences); store wrapped in aluminum foil.

2.2. Genomic DNA Extraction

1. Cell lysis solution: 0.5% SDS, 10 mM Tris-HCl pH 8.0, 300 mM NaCl, 5 mM EDTA, and 200 μg/ml Proteinase K.

2. Phenol/chloroform/isoamyl-alcohol.

3. Cholorform/isoamyl-alcohol.

4. DNase-free ribonuclease (RNase A).

5. Ethanol, 100% and 70%.

2.3. Purification of Heavy/Light DNA by CsCl Density Centrifugation

1. Restriction endonucleases *EcoR*I and *Hind*III (NEB).

2. CsCl solution: Use 1 g CsCl (Sigma) per ml of TE buffer (10 mM Tris-HCl pH 8.0, 1 mM EDTA pH 8.0). After the CsCl is dissolved in solution, the refractive index should be 1.4052 (*see* **Note 1**).

3. OptiSeal tubes (Beckman).

4. Quick-Seal tubes (Beckman).

2.4. BrdU ELISA

1. 2X SE: 0.8 M NaOH, 20 mM EDTA.

2. 96-well ELISA plate (BD BioCoat™).

3. 2 M Ammonium acetate (pH 7.0).

4. Non-fat dried milk.

5. Anti-BrdU (Monoclonal antibody to the thymidine-analogue 5-bromo-2'-deoxyuridine Fab fragments with peroxidase (POD) conjugated; Roche Applied Sciences).

6. TMB substrate (PIERCE).

7. 2 M H_2SO_4.

2.5. Fragmentation and Labeling of H/L DNA

1. DNaseI (Epicenter).

2. 10X One-Phor-All buffer (Amersham-Pharmacia).

3. 5X TdT buffer (Roche).

4. 25 mM $CoCl_2$ (Roche).

5. 1 mM bio-ddATP (Enzo Life Sciences).

6. Terminal deoxytransferase (400 U/ml; Roche).

2.6. Genome-Tiling Array Hybridization

Microarrays: To generate replication time profiles ENCODE01-Forward (Affymetrix, Santa Clara, CA) tiling arrays were used. These arrays are designed to study the pilot ENCODE regions of DNA, comprised of 30 Mb of DNA, or approximately 1% of the human genome. These pilot regions were selected by a committee of the National Human Genome Research Institute (NHGRI). Half of the content on the ENCODE01 Array was manually selected by the NHGRI committee, while the remaining 50% were randomly selected *(8)*. A total of 14.82 Mb of sequence constituted the manually selected regions and included 14 target regions ranging in size from 500 kb to 2 Mb. The randomly selected content includes 30, 500 kb regions selected based on gene density and level of non-exonic conservation.

3. Methods

To generate temporal maps of replication, cell populations are synchronized at G1/S by thymidine/aphidicolin block and released into S-phase as described in the following section. After release from block, replicating DNA is pulse-labeled with 5-bromo-2′-deoxyuridine (BrdU) for successive 2 h intervals of S-phase, thereby dividing 10 h of S-phase into five time intervals (**Fig. 14.1B**). The efficiency of the block and release is ascertained by propidium iodide based FACS (flow cytometry) (**Fig. 14.1A**). Genomic DNA is isolated from all five time intervals (0–2, 2–4, 4–6, 6–8, and 8–10 h) representing 10 h of the entire S-phase (**Fig. 14.1B**). Genomic DNA is digested with *Eco*RI/*Hind*III restriction enzymes (**Fig. 14.1C**) and then the BrdU-incorporated heavy/light (H/L) DNA is purified by CsCl density gradient (**Fig. 14.1D**). Success of purification is assayed by BrdU ELISA (**Fig. 14.2**). Purified DNA from each time interval is fragmented to 50–200 bp by DNaseI as described later. The DNA fragments are end-labeled with biotinylated-ddATP using terminal transferase (**Fig. 14.1E**). This labeled DNA is then hybridized to the high-density genome-tiling Affymetrix array.

During computational processing of the data, it is important to identify two separate classes of replication, temporally specific replication (TSR) and temporally non-specific replication (TNSR). An area undergoes TSR when all alleles at that locus replicate synchronously in S-phase. A non-trivial portion of a given genome may replicate alleles in a non-synchronous manner, with some replicating early and others later. For a given probe on the array, the algorithm to classify it as temporally specific versus non-specific considers the signal of that probe across all time points. If there is evidence that significant replication is occurring in two non-

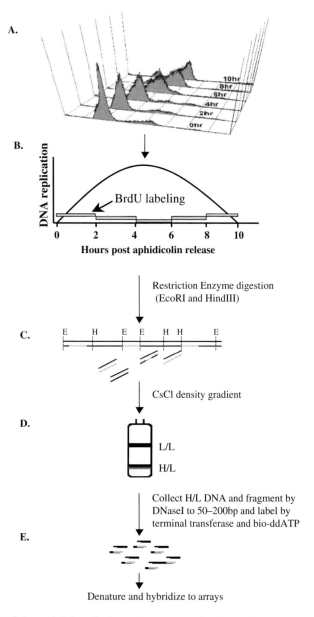

Fig. 14.1. **Schematic of methods used to map replication timing.** (**A**) Synchronous progression of HeLa cell through S-phase and harvesting of timed replication pools. HeLa cells released from a G1-S block followed by FACS for DNA content (*(9)*; "Copyright (2005) National Academy of Sciences, USA"). (**B**) Cells synchronously released in the same way as in (**A**) were pulsed with BrdU at indicated time intervals. (**C**) Isolation of DNA (BrdU substituted; in *gray*) and digestion to produce fragments that resolve well H/L and L/L DNA. (**D**) *Top black* band represents LL DNA while the *lower gray/black* band represents H/L DNA. (**E**) Purified H/L DNA was fragmented with DNaseI and labeled with biotin-ddATP by terminal transferase and hybridized to the ENCODE arrays.

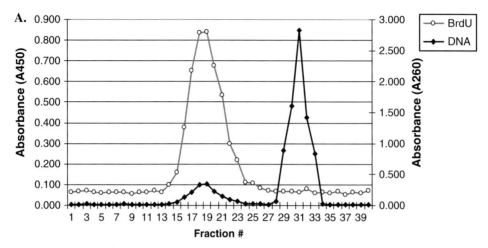

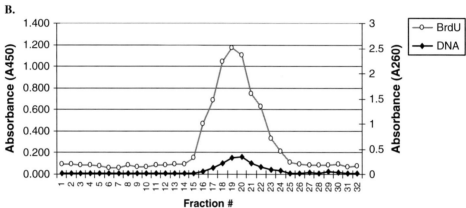

Fig. 14.2. **BrdU ELISA to check fractionation of H/L DNA**. (**A**) Genomic DNA labeled during 0–2 h of S-phase. *Gray line* is the Anti-BrdU reactivity in an ELISA (measured as absorbance at 450 nm) and *black line* is the absorbance at 260 nm for total DNA. The H/L DNA forms a peak away from the bulk of the unreplicated DNA. (**B**) Second purification of H/L fractions. Complete overlap of BrdU peak with total DNA peak indicates purity of H/L DNA.

adjacent time points, then the probe is classified as temporally non-specific. In the absence of such evidence, where significant replication is isolated to a single time point or two adjacent time points, the probe is classified as temporally specific. Later, broad regions of TSR and TNSR are segregated using a majority algorithm where the ratio of temporally specific to temporally non-specific probes is considered.

3.1. Cell Synchronization and Pulse Labeling

1. Treat 10–30 × 15 cm plates (10 plates for 2–4 h, 4–6 h, and 6–8 h time points and 30 plates each for 0–2 h and 8–10 h time points) of cells at 60% confluency with 2 mM thymidine for 12 h.

2. Release cells from thymidine block by removing the media and washing three times with PBS.

3. Add 20 ml of fresh media and incubate for 12 h.

4. Add 1 μg/ml aphidicolin and incubate for 12 h.

5. Release the cells from aphidicolin block by removing the media and washing three times with PBS.

6. Add 20 ml of fresh media to release the cells into S-phase of cell cycle. Label cells with 100 μM BrdU for 2 h.

7. Remove the media and wash cells thrice with PBS and then trypsinize them.

8. Harvest the cells by centrifugation at 200g for 5 min in Eppendorf centrifuge 5810 (or equivalent). Save 5 × 10^5 cells for FACS (fix cells in 70% ethanol for at least 1 h at 4°C. Stain in 1 ml propidium iodide solution for 1 h).

9. Proceed to genomic DNA extraction step. If the genomic DNA extraction is not to be performed the same day, freeze the cell pellets at −80°C.

3.2. Genomic DNA Extraction

1. Collect 10^8 cells (monolayer cells) in a 15 ml tube. Then centrifuge at 200g for 5 min.

2. Wash 2X with 10 ml PBS.

3. Add 10 ml cell lysis buffer to cell pellet and resuspend the cells by gently tapping the tube with finger and incubate at 55°C for 2 h.

4. Add 10 ml of Phenol/Chloroform/IAA (25:24:1), rotate 10 min at room temperature. Centrifuge at 1,731g for 10 min, transfer supernatant to a new tube.

5. Add equal volume of chloroform/isoamyl alcohol (24:1). Rotate 10 min at room temperature. Centrifuge at 1,731g for 10 min, and transfer the supernatant to a new tube.

6. Add DNase-free ribonuclease A (RNase A) to a final concentration of 25 μg/ml and incubate for 1 h at 37°C.

7. Repeat Steps 4 and 5 and transfer the aqueous phase to SS-34 rotor Sorvall tubes.

8. Add 2 volumes ethanol, mix by inverting the tube. Leave on ice for 10 min.

9. Centrifuge at 12,000g at room temperature for 30 min in Sorvall RC5-B centrifuge (or equivalent tubes/rotor). Decant the supernatant by gently inverting it.

10. Wash the pellet with 10 ml 70% ethanol. Centrifuge at 12,000g for 10 min in RC5-B centrifuge.

11. Decant the supernatant by gently inverting it.

12. Air dry the pellet for about 5–10 min (*see* **Note 2**).

13. Resuspend the pellet in 1 ml sterile water and incubate at 37°C for an hour till DNA is completely dissolved.

14. Record concentration and yield by reading the absorbance at 260 nm.

3.3. H/L DNA Purification by CsCl Gradient

1. Digest genomic DNA with 5 U/μg each of *Eco*RI and *Hin*-*d*III restriction enzymes for 5 h. Add EDTA to the reaction at 1 mM concentration.

2. Check for completion of digestion by running 10 μl of the reaction volume on a gel. Typically, the spread of digested DNA will be more abundant in the 2–5 kb range. If this is not the case then add more enzyme and leave it longer for digestion.

3. Prepare 1 g/ml CsCl solution in TE buffer pH 8.0 (Tris-HCl 10 mM, EDTA 1 mM).

4. Add 1 g CsCl to the digested DNA (400 μg) and make up the volume to 1 ml (*see* **Note 3**).

5. Carefully pour DNA-CsCl solution into OptiSeal tubes. Fill rest of the tube with CsCl solution.

6. Centrifuge the density gradients at 25°C (*see* **Note 2**) in Beckman VTi50 rotor (or equivalent tubes/rotor) at 167,200*g* for 48 h, no brake (i.e., deceleration = 0).

7. After ultracentrifugation, carefully collect fractions at flow rate of 1 ml/min, starting from bottom of the tube.

8. Check for the H/L peak by performing BrdU ELISA (**Fig. 14.2A**).

9. If density banding worked fine then pool the fractions representing H/L peak in Quick-Seal tubes and ultracentrifuge again using NVT90 rotor (or equivalent tubes/rotor) for 18 h at 292,200*g*, no brake (i.e., deceleration = 0).

10. Collect the fractions at flow rate of 200 μl/min from bottom and perform BrdU ELISA to ascertain separation of H/L DNA (**Fig. 14.2B**). This second centrifugation step further cleans up the H/L DNA from any contaminating LL DNA.

11. Pool the fractions representing H/L DNA, dialyze for 5 h (or overnight) in TE buffer.

3.4. BrdU ELISA

1. Spot 2.5 μl H/L DNA + 12.5 μl water + 15 μl of 2X SE on poly-lysine 96-well ELISA plate. Make sure to have no-DNA control wells to evaluate the background. Also perform the whole assay in triplicate.

2. Denature DNA for 10 min on 100°C hot plate.

3. Add 30 μl of 2 M ammonium acetate (pH 7.0) and 40 μl of TE (pH 8.0) to each well and leave it on shaker for 30 min at RT.

4. Change the solution with 120 μl each of 5% non-fat milk in PBS containing 10 mM EDTA and shake for 30 min at RT.

5. Rinse the wells three times with 150 μl each of PBS.

6. Prepare 1:100 dilution of Anti-BrdU in PBS. Add 100 μl in each well and shake the plate at RT for 30 min (*see* **Note 5**).

7. Wash the plates three times with 150 μl of PBS. Each wash should be 10 min with shaking at RT.

8. Develop the color (blue) using 100 μl of TMB substrate. Developing time can vary from 1 min to 30 min (*see* **Note 6**).

9. Quench with 100 μl of 2 M H_2SO_4 (color turns yellow).

10. Read the absorbance at 450 nm.

3.5. Fragmentation and Labeling of H/L DNA

1. Dilute DNaseI 1:16 with 1X One-Phor-All buffer and keep it on ice. Do not vortex any solutions containing DNaseI.

2. Set up the reaction in 40 μl as detailed in the **Table 14.1**.

Table 14.1
Conditions for setting up DNaseI digestion of H/L DNA

Reagents	Final concentraion or amount in reaction	Reaction conditions
10X One-Phor-All buffer	4 μl	37°C for 4 min and then hold on ice
DNase diluted 1:16 with 1X One-Phor-All buffer	5 μl	
DNA	12 μl	
Water	To 40 μl	
Total volume	**40 μl**	

3. Check on 2% agarose gel, the bulk DNA should be 50–100 bp. If not then incubate the tube again at 37°C (*see* **Note 7**).

4. If the digestion is complete to desired fragment length then inactivate DNaseI at 99°C for 10 min.

5. For labeling assemble the reaction as in **Table 14.2**.

6. The labeled (ds) DNA is now ready to be hybridized to the array. Store labeled DNA at −80°C till further use.

3.6. Hybridization and Washing of Microarrays

1. Hybridize, wash, and stain the chips as per manufacturer's protocol (FS450_0001).

2. Scan and analyze each microarray for signal intensities using GeneChIP® Scanner 3000 and GeneChIP Operating Software (GCOS) from Affymetrix.

Table 14.2
Conditions for labeling of H/L DNA using terminal transferase

Reagents	Volume	Reaction conditions
DNA from fragmentation	34 μl (~9 μg)	Incubate 37°C for 2 h
5X TdT buffer	14 μl	
25 mM CoCl$_2$	14 μl	
1 mM bio-ddATP	5 μl	
Terminal deoxytransferase (400 U/ml)	3 μl	
Total volume	70 μl	

3.7. Time of Replication of 50% (TR50) Calculation

1. Calculate a signal value for each probe of each array as Max [PM (Perfect Match) – MM (Mis Match), 0]. This means that if PM–MM is <0, it is set as 0.

2. Calculate a TR50 value for each probe of the array set by linearly interpolating the point at which 50% of signal is accumulated across all time points. Any probe of the array set that shows 0 signal in every time point is excluded from further analysis.

3. Classify each probe of the array set as TSR or TNSR by the following criteria:

   a. Find the minimum signal of all time points, call it MIN.

   b. Subtract MIN from each time point to adjust their signal.

   c. Call the number of expected alleles at each locus of the cell line in question N.

   d. Sum the total signal across all time points, call it TOTAL.

   e. Find the maximum signal value of all time points, call it MAX.

   f. Find the maximum sum of all sets of two adjacent time points, call it MAXSUM.

   g. Find the maximum sum of all sets of two adjacent time points where neither time point in the sum has signal equal to MAX, call it MAXSUMNOT.

   h. If MAXSUM > $(1-1/N)$ * Total, then classify the probe as TSR.

   i. Otherwise, if MAXSUMNOT ≥ $(1/N)$ * Total, then classify the probe as TNSR.

   j. Otherwise, if both Steps h and i fail to classify the probe, then classify it as TSR.

3.8. Segregation of TSR and Pan-S Regions

1. Process the probes with a sliding window of 60,000 bp. When the number of probes in the window is ≥ 600, generate an interval. If the number of TSR probes is greater than the number of TNSR probes, generate a TSR interval. If the number of TSR probes is less than the number of TNSR probes, generate a Pan-S interval. If they are equal, use the next probe to break the tie. Whenever the ratio changes, switch to generating the opposite interval. If the number of probes in the window drops below 600, end the current interval.

2. The mutually exclusive TSR intervals and Pan-S intervals divide the area on the array with the required minimum probe density into TSR regions and Pan-S regions. Remove intervals with length less than 10,000 bp. Then within each set, join intervals whose endpoints are less than 10,000 bp apart.

3.9. TR50 Smoothing

1. Perform a Lowess smoothing on the set of TSR probes TR50 values with the smoother set to consider a window of 60,000 bp.

2. The resulting smoothed curve is called the smoothed TR50 and is paired with the segregation intervals to comprise the replication profile.

3.10. Discrete Temporally Specific Timing Categories

1. Collect the set of smoothed TR50 probe values that fall into TSR regions.

2. Sort the smoothed TR50 values of these probes.

3. Choose the one-third point of the distribution as the early-mid cutoff and the two-thirds point of the distribution as the mid-late cutoff.

4. Break the TSR regions up into subregions based on the smoothed TR50 values. A smoothed TR50 value less than the early-mid cutoff gives rise to an early interval. A smoothed TR50 value less than the mid-late cutoff but later than the early-mid cutoff gives rise to a mid interval. A smoothed TR50 value greater than the mid-late cutoff gives rise to a late interval.

5. Remove intervals with length less than 10,000 bp from each of these sets: early intervals, mid intervals, and late intervals.

6. Join intervals in each set that are separated by less than 10,000 bp.

7. The final replication timing profile is attained by pairing the smoothed TR50 curve with the discrete non-overlapping segregation intervals: early, mid, late, and Pan-S. **Figure 14.3** shows an example of the replication timing profile for part of human chromosome 21 in HeLa cell line.

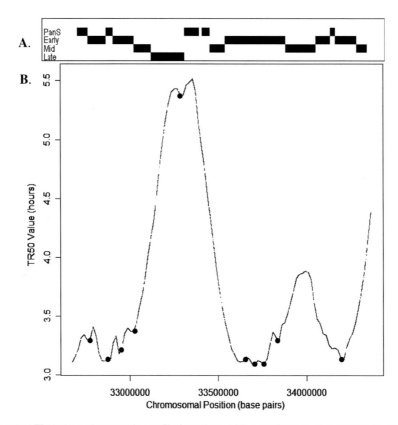

Fig. 14.3. Smoothed TR50 plot and segregation profile from the 1.9 Mb part of human chromosome 21 in HeLa cell line. (A) Segregation of replication timing profile into early, middle, late, and Pan-S replicating regions along the length of the chromosome. (B) TR50 plot. The lowest point (*line plot*) in each valley indicates a site that is replicated before its adjoining segments and so is likely to contain origins of replication. The *solid circles* represent these putative origins. The *gaps* in the TR50 plots indicate the presence of repeats. In order to minimize cross-hybridization of oligonucleotides, repeat regions of the genome are not spotted on the tiling arrays.

4. Notes

1. Before estimating the refractive index of the CsCl solution, calibrate the instrument with distilled water (refractive index of 1.3330).

2. Do not dry the DNA completely: dessicated DNA is very difficult to dissolve. To get genomic DNA in solution, first dissolve at 37°C for at least an hour and then leave at 4°C overnight.

3. Do not forget to set refractive index of the DNA with CsCl. It should be 1.4052.

4. Always run the CsCl gradients at RT (25°C). Running the gradient at lower temperature can cause crystallization of CsCl, which will affect the gradient and can also be detrimental to rotor and the ultracentrifuge.

5. We have found that Anti-BrdU-POD antibody to be excellent for this assay. It also negates the use of secondary antibody.

6. Do not let the color turn dark blue after addition of TMB substrate and green after addition of H_2SO_4: this is the sign of signal saturation and hence can give misleading results.

7. DNaseI digestion: Since efficiency of DNaseI digestion is sensitive to a number of conditions (e.g., purity of DNA, lot and vendor of DNaseI) it is necessary to titrate the DNaseI amount. The conditions mentioned here therefore should only be considered as guidelines that may have to be adjusted for a particular DNAseI prep.

Acknowledgments

This work was supported by National Institutes of Health Grant HG003157 (to A.D.)

References

1. Bell, S.P. and Dutta, A. (2002) DNA replication in eukaryotic cells. *Annu Rev Biochem*, **71**, 333–374.

2. Watanabe, Y., Fujiyama, A., Ichiba, Y., Hattori, M., Yada, T., Sakaki, Y. and Ikemura, T. (2002) Chromosome-wide assessment of replication timing for human chromosomes 11q and 21q: disease-related genes in timing-switch regions. *Hum Mol Genet*, **11**, 13–21.

3. Sinnett, D., Flint, A. and Lalande, M. (1993) Determination of DNA replication kinetics in synchronized human cells using a PCR-based assay. *Nucleic Acids Res*, **21**, 3227–3232.

4. Selig, S., Okumura, K., Ward, D.C. and Cedar, H. (1992) Delineation of DNA replication time zones by fluorescence in situ hybridization. *Embo J*, **11**, 1217–1225.

5. Woodfine, K., Fiegler, H., Beare, D.M., Collins, J.E., McCann, O.T., Young, B.D., Debernardi, S., Mott, R., Dunham, I. and Carter, N.P. (2004) Replication timing of the human genome. *Hum Mol Genet*, **13**, 191–202.

6. Karnani, N., Taylor, C., Malhotra, A. and Dutta, A. (2007) Pan-S replication patterns and chromosomal domains defined by genome-tiling arrays of ENCODE genomic areas. *Genome Res*, **17**, 865–876.

7. Raghuraman, M.K., Winzeler, E.A., Collingwood, D., Hunt, S., Wodicka, L., Conway, A., Lockhart, D.J., Davis, R.W., Brewer, B.J. and Fangman, W.L. (2001) Replication dynamics of the yeast genome. *Science*, **294**, 115–121.

8. ENCODE project consortium. (2004) The ENCODE (ENCyclopedia Of DNA Elements) Project. *Science*, **306**, 636–640.

9. Jeon, Y., Bekiranov, S., Karnani, N., Kapranov, P., Ghosh, S., MacAlpine, D., Lee, C., Hwang, D.S., Gingeras, T.R. and Dutta, A. (2005) Temporal profile of replication of human chromosomes. *Proc Natl Acad Sci USA*, **102**, 6419–6424.

Chapter 15

Integration of Diverse Microarray Data Types

Keyan Salari and Jonathan R. Pollack

Abstract

Over the past decade, DNA microarrays have proven to be a powerful tool in biological research for the molecular surveillance of cells and tissues. The expansive utility of DNA microarrays owes its nascence to the development of a multitude of microarray platforms that enable the systematic and comprehensive exploration of diverse genomic properties and processes. Concomitant with the explosive generation of microarray data over the last several years has been an increasing interest in the integration of such diverse data types, thus spurring the development of novel statistical techniques and integrative bioinformatics tools. This chapter will outline general approaches to microarray data integration and provide an introduction to DR-Integrator, a broadly useful analysis tool for the integration of DNA copy number and gene-expression microarray data.

Key words: Data integration, integrative genomics, DR-Integrator, array CGH, gene-expression.

1. Introduction

The first application of DNA microarray technology was for measuring genome-wide mRNA transcript levels, commonly referred to as gene-expression profiling *(1)*. While still remaining the most popular use of the technology, a host of new microarray platforms have since been developed that explore a wide range of properties and processes of the cell. Among other applications, microarrays have been adapted to generate genome-scale data on microRNA expression *(2)*, protein expression *(3)*, DNA copy number alterations (CNAs) *(4, 5)*, variation among single-nucleotide polymorphisms (SNPs) *(6)*, DNA methylation *(7)*, physical interactions between chromatin and transcriptional regulators *(8)*, and RNAi-based loss-of-function phenotypes *(9)*. While each microarray data type offers a unique snapshot

Jonathan R. Pollack (ed.), *Microarray Analysis of the Physical Genome: Methods and Protocols, vol. 556*
© Humana Press, a part of Springer Science+Business Media, LLC 2009
DOI 10.1007/978-1-60327-192-9_15 Springerprotocols.com

of a cell's state, an analysis integrating information from two or more of these complementary data types (termed "integrative genomics") can reveal much more than the sum of its parts. As such, many investigators have innovated and developed novel statistical methods and tools to integrate microarray data across different platforms, data types, and laboratories.

1.1. Single Data Type Integration

Within a single microarray data type, such as gene expression, integrating multiple datasets increases statistical power and allows for a more comprehensive study of the commonalities and variations characterizing the samples profiled. In one exemplary study, Segal et al. integrated data from 1,975 published microarray experiments spanning 22 tumor types, wherein the authors characterized tumors as a combination of activated and deactivated modules (sets of genes that act in concert to carry out a specific function) (10). The authors describe activation of some modules as specific to particular types of tumors while other modules were shared across a diverse set of clinical conditions. Underscoring the importance of making published microarray data publicly available, the collective series of published microarray experiments included in this integrative study provided insights into the biology of human cancer that any single dataset alone would not have been able to reveal.

1.2. Multiple Data Type Integration

While each microarray data type alone provides important information, integration across multiple layers of genome-scale data (e.g., DNA CNAs and gene-expression levels) can reveal additional insights. Adding each layer of data not only provides another complementary view of the molecular profile of a sample, but it also allows for the study of the "interactome", or the network of interactions and regulatory relationships between all cellular molecules (11). For example, SNP genotype data and gene-expression data can be combined to study the effect of genetic variation on gene expression (12). Other combinations of genome-scale data can also be integrated to provide novel biological insights, such as data from chromatin immunoprecipitation followed by microarray analysis (ChIP-chip) with gene-expression data (13, 14), DNA copy number with gene-expression data (15–18), DNA copy number with epigenomic data (19), and genomic data between different species (an approach that has been dubbed "comparative oncogenomics") (20–22).

While the aim of each integrative genomics study may differ, the approaches to data integration can be abstracted to formulate a general framework for analysis. Importantly, additional considerations must be made to properly integrate datasets of diverse data types, beyond the challenges involving data normalization and interplatform comparability addressed in single data type integration. As a guiding principle for multiple data type integration, three aspects of the datasets should be evaluated and should inform the approach of an integrative analysis: the context, catalog, and content (23).

The experimental "context", or conditions under which the data were measured, should ideally be identical (e.g., breast carcinoma DNA copy number data and breast carcinoma gene-expression data). The most preferable experimental design subjects the same biologic samples to analysis by each microarray platform so that the data are paired. Although methods exist to integrate data collected from different patients (i.e., unpaired data) *(24)*, patient-to-patient molecular variation is sufficiently large that directly integrating data from the same individual is much more powerful and efficient than comparing the averages between the data types.

Next, the "catalog" (the set of elements/molecules measured) for each microarray platform should be evaluated for comparability. Each microarray data type to be integrated is likely to have been generated using a different platform (e.g., Agilent 44 K oligonucleotide arrays for measurements of DNA CNAs and Affymetrix U133 GeneChip arrays for gene-expression measurements). Therefore, an element-to-element mapping scheme should be established between the catalogs of each microarray platform. One approach is to translate each dataset to a universal catalog by mapping probes from each platform to each other using a common identifier (e.g., Entrez Gene ID, UniGene Cluster ID, or gene symbol). In the example case of the Agilent and Affymetrix datasets, each set of Agilent probes interrogating genomic regions within a single gene could be averaged, while the Affymetrix probe sets interrogating the same game could be averaged. The resulting dataset would thus contain an average DNA copy number and gene-expression measurement for each gene, and the set of genes measured by both microarray platforms would be used in the integrative analysis. Of note, translating catalogs of different microarray platforms often requires prior biological knowledge (e.g., homology of genes between species, mapping of genes to proteins, etc.), and several resources containing such information are available from the National Center for Biotechnology Information (NCBI; http://www.ncbi.nlm.nih.gov/).

Finally, the microarray data "content" (the actual data measurements; e.g., 1-channel signal intensities vs. 2-channel $\log_2$ ratios) should be considered to determine the most appropriate method for quantitative or qualitative integration. Some microarray platforms provide absolute quantitative measurements while others provide relative quantitative measurements (e.g., Affymetrix arrays measure absolute transcript abundance, while two-channel platforms typically measure transcript abundance in a test sample relative to a common reference). Still others provide a quantitative measurement (i.e., signal intensity) of an underlying binary event, such as whether or not a genomic region is methylated or is bound by a transcription factor of interest. Appropriate data transformations should be made, if possible, to prepare datasets with equivalent content. For example, Affymetrix single-channel data can be transformed into $\log_2$ ratios (typical of two-channel arrays) by dividing each probe set's average intensity by the average intensity for that probe set across all the samples, and then log-transforming these ratios. The statistical technique employed for integration should account for the absolute vs. relative nature of the quantitative measurements. For data representing binary events, a threshold should be established to assign each continuous measurement a discrete value (e.g., methylated vs. not methylated), and simple intersection of gene lists may in such cases be the most appropriate method for integration with other microarray data types.

1.3. Integrating DNA Copy Number and Gene-Expression Data for Cancer Gene Discovery

A focus of many integrative genomics studies has been cancer, among other diseases. Cancer is a genetic disease, resulting from the accumulation of alterations or mutations in key genes regulating processes like cell proliferation and cell death. Together, these processes drive cancer development and progression. DNA CNA represents one form of genetic alteration that has been extensively characterized in a variety of tumor types using array-based comparative genomic hybridization (array CGH). CNAs lead to the amplification and deletion of oncogenes and tumor-suppressor genes (TSGs), respectively, and thereby play a critical role in tumorigenesis. While delineating CNAs across many samples facilitates the identification of oncogenes (in regions of recurrent amplification) and TSGs (in regions of recurrent deletion), cumulatively such genetic changes often span a substantial proportion of the genome, thereby obfuscating the distinction between "driver" cancer genes selected for by a genetic event and nearby "passenger" genes incidentally co-amplified or deleted. Analogously, when comparing cancer cells to normal cells, thousands of genes are often differentially expressed, rendering discrimination of the most salient, primary changes from correlated, downstream changes difficult.

One useful approach to aid cancer gene discovery is to integrate DNA copy number and gene-expression profiles (16–18, 25). Tumors often harbor CNAs altering the gene dosage of hundreds or thousands of genes. However, tissue-specific expression or feedback regulation, among other mechanisms, may leave expression levels of many of these genes unaltered. Because the effects of CNAs are mediated by changes in gene expression, the subset of genes exhibiting both concordant changes in DNA copy number and gene-expression (e.g., amplified and overexpressed genes) are likely to be enriched for candidate oncogenes and TSGs. Consequently, there has been increasing interest in integrating array CGH and gene-expression microarray data for cancer gene discovery, and in developing novel statistical techniques to facilitate such integrative analyses.

The remainder of this chapter will focus on the practical issues of integrating paired DNA copy number and gene-expression data, and will demonstrate one approach to an integrative analysis using the software tool DR-Integrator (*DNA/RNA-Integrator*) (26).

2. Data Preprocessing

Integrating DNA copy number and gene-expression data is most powerful when such data measurements are paired; i.e., the experimental contexts of each data type are identical, as the same samples are profiled by both CGH and gene-expression microarrays. Before an integrative analysis can be performed, each dataset must be normalized and then merged. The paired data measurements, merged into one DNA/RNA dataset, can then be analyzed using DR-Integrator (**Fig. 15.1**).

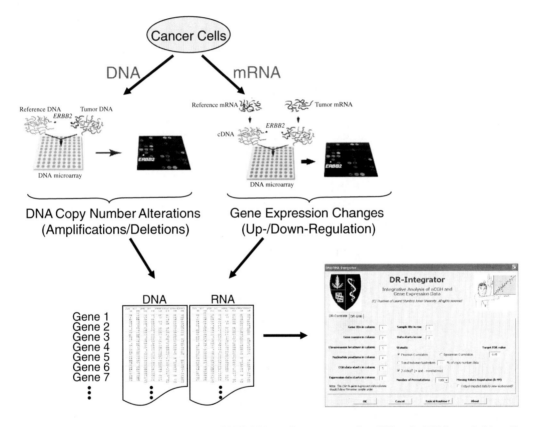

Fig. 15.1. **Schematic of paired measurement of DNA CNAs and gene expression.** DNA and mRNA harvested from the same biologic specimen are fluorescently labeled and hybridized to a microarray for genome-wide copy number and gene-expression analysis. The datasets are then processed, merged, and imported to Microsoft Excel for analysis with DR-Integrator.

2.1. Preprocessing/ Normalizing DNA Copy Number Data

DNA copy number data should be preprocessed and normalized as would be done in a conventional single data type analysis. Briefly, for data derived from two-color microarrays, background-subtracted $\log_2$ fluorescence ratios should be normalized for each array by mean centering (so that the average of all $\log_2$ ratios is set to 0). Probes with fluorescence intensities in the Cy3 reference channel at least 1.5-fold above background should be considered well-measured and used for subsequent analysis. For data derived from SNP arrays, manufacturer-specific software such as Affymetrix MAS or Genotyping Console Software should be utilized to filter out poorly performing probe sets (*see* **Note 1**). Each probe (oligonucleotide or cDNA) should be assigned a map position using the latest NCBI genome assembly, which can be accessed through the UCSC genome browser database. This information may also be provided by the microarray manufacturer. The average log_2 ratio should be used for genes represented by multiple arrayed probes. In order to identify genomic regions with significant DNA copy number changes, the dataset should be run through an algorithm that objectively calls significant gains and losses of chromosomal regions based on a user-defined false discovery rate (FDR) (typically 5%). The fused lasso regression method *(27)*, available under the R package name "cghFLasso" at http://cran.r-project.org/, is a robust and computationally efficient method for calling gains and losses (*see* **Note 2**).

2.2. Preprocessing/ Normalizing mRNA Expression Data

Gene-expression data should be preprocessed and normalized as would be done in a conventional single data type analysis. Briefly, for two-color microarray data, background-subtracted fluorescence log_2 ratios should be globally normalized for each array, and then mean-centered for each gene (i.e., reporting expression relative to the average log_2 ratio across all samples). Genes with fluorescence intensities in the Cy5 or Cy3 channel at least 1.5-fold above background in at least half of the samples should be considered well-measured and included for subsequent analysis. For Affymetrix gene-expression arrays, MAS or Expression Console Software can be used to perform probe set normalization and filtering (*see* **Note 1**). Before merging the DNA copy number and gene-expression datasets, the content of each data type should be evaluated for comparability (e.g., both expressed as log_2 ratios).

2.3. Merging DNA/RNA Datasets

The objective of merging the DNA copy number and gene-expression datasets is to obtain a copy number and transcript level measurement for each gene interrogated by both microarray platforms. In other words, each platform's catalog must be compared to pair the measurements made for the same gene. One simple approach is to assign, for each gene with a measurement from the expression array, a DNA copy number from either a probe interrogating the same named gene (or gene ID), or the average DNA copy number of the nearest 5′ and 3′ probes. Because CNAs are discrete genetic events usually spanning more than a single gene, all probes interrogating the same CNA should theoretically have the same DNA copy number. Thus, averaging the copy number from neighboring probes is a reasonable surrogate when a probe interrogating the exact gene region is unavailable. The final dataset should be a matrix of m rows and $2n$ columns, where m is the number of genes with both DNA copy number and gene-expression measurements, and n is the number of samples profiled on each microarray platform. Columns 1 to n should contain the DNA copy number data, and columns $n + 1$ to $2n$ should contain the gene-expression data, with the samples ordered identically.

3. Methods

The DR-Integrator (DNA/RNA-Integrator) software contains two analysis tools: (1) DR-Correlate: an analysis that identifies genes with significant correlations between their DNA copy number and gene-expression; and (2) DR-SAM: a supervised analysis to identify genes with significant alterations in both DNA copy number and gene expression between two different sample classes (e.g., tumor subtypes). DR-Integrator works as a user-friendly plug-in application to Microsoft Excel, and is freely available from the Pollack Lab (http://pollacklab.stanford.edu/) (*see* **Note 3**). The underlying statistical methods are implemented in the R programming language, and are also available under the package name "DRI" at http://cran.r-project.org/. The final dataset generated in **Section 2.3** can be imported directly into a Microsoft Excel spreadsheet for analysis by DR-Integrator.

3.1. DR-Correlate: Statistical Approach

This tool performs an analysis to identify all genes with statistically significant correlations between their DNA copy number and gene expression. In other words, DR-Correlate identifies genes with expression changes explained by underlying DNA CNAs. Three statistics for measuring correlation are implemented: (1) Pearson's correlation; (2) Spearman's correlation; and (3) an "extreme t-test". For Pearson's and Spearman's correlations, the respective correlation coefficient is computed for each gene. For the extreme t-test, a modified Student's t-test is computed for each gene, comparing gene-expression levels of samples from the lowest and the highest deciles with respect to DNA copy number. In other words, for each gene the samples are rank-ordered by DNA copy number and samples below the 10th percentile and above the 90th percentile form the two groups whose gene expression is tested for a significant difference using a t-statistic. The percentile threshold defining the two groups is user-defined.

Because a correlation statistic is computed for a large number of genes, assessing the statistical significance of the test scores must account for multiple hypothesis testing. In the analysis of genome-scale data, this is most often achieved by FDR estimation. In this approach, one first generates a series of randomized datasets equivalent to the original dataset with the exception that the hypothesized associations are disrupted. In the case of our DNA/RNA data matrix with identically ordered samples (or columns), randomly permuting the sample labels will disrupt the correlations between the paired DNA copy number and gene-expression measurements. For each randomized data matrix, the correlation statistic is computed for every gene. The correlation statistics derived from each randomized dataset collectively constitute a well-sampled null distribution. The observed correlation scores from the original dataset are then compared to this null distribution to estimate a FDR, signifying for a given correlation score, the number of genes expected by chance to have an equivalent or greater score.

3.2. DR-SAM: Statistical Approach

DR-SAM (DNA/RNA-Significance Analysis of Microarrays) performs a supervised analysis to identify genes with statistically significant differences in both DNA copy number and gene-expression between different classes (e.g., tumor subtype-A vs. tumor subtype-B). The goal of this analysis is to identify genetic differences (CNAs) that mediate gene-expression differences between two groups of interest. DR-SAM implements a modified Student's t-test to generate for each gene two t-scores assessing differences in DNA copy number and differences in gene expression. A final score (S) is computed by first summing the copy number t-score and gene-expression t-score, and then weighting the sum by the ratio of the two t-scores ($0 \leq w \leq 1$) The weight is applied to favor genes with strong differences in both DNA copy number and gene-expression between the two classes; i.e., a gene with statistically equal differences in copy number and in gene-expression (i.e., copy number t-score = gene-expression t-score) will have a weight of 1, while genes with unequal contributions from copy number and expression will have a weight less than 1, resulting in a lower score.

$$S = w^{*}(t_{DNA} + t_{RNA})$$

$$\text{where } w = \min\left\{ \frac{t_{DNA}}{t_{RNA}}, \frac{t_{RNA}}{t_{DNA}} \right\}$$

Statistical significance of the DR-SAM scores, accounting for multiple hypothesis testing, is assessed by randomly permuting the data to estimate a FDR, as described for the DR-Correlate analysis (*see* **Section 3.1**).

3.3. Usage

To run either the DR-Correlate or DR-SAM analysis in Microsoft Excel, select the cells containing the entire DNA/RNA dataset and click the "DR-Integrator" button on the toolbar. A dialog box will appear, and the tab corresponding to the desired analysis tool

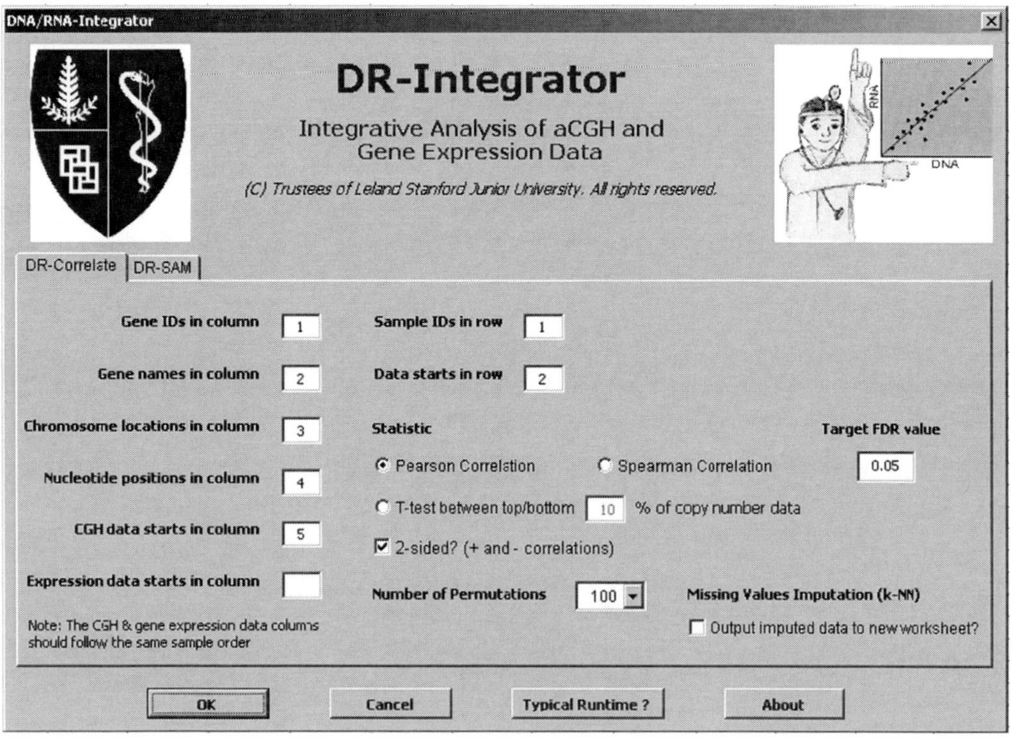

Fig. 15.2. **DR-Integrator user interface.** Shown is the dialog box in Microsoft Excel for DR-Integrator. The appropriate tab for either DR-Correlate or DR-SAM analysis should be selected, and the appropriate data specifications and user options should be filled in.

should be selected (**Fig. 15.2**). Text boxes specifying the columns corresponding to gene identifiers, gene names, chromosomal locations, nucleotide positions, DNA copy number data, and gene-expression data should be filled appropriately (note: the sample order among the copy number data should be identical to the sample order in the gene-expression data). For DR-SAM analysis, fill in the appropriate text box specifying the row containing the group labels (1 vs. 2). The test statistic of choice should be selected (Pearson, Spearman, or extreme *t*-test for DR-Correlate) (*see* **Note 4**); if extreme *t*-test is selected, fill in the desired percentile threshold to establish low and high copy number groups (default: 10%). If desired, select the box to output a new worksheet with missing data values filled in by imputation (*see* **Note 5**). Select the number of permutations to perform to generate the randomized datasets (default: 1,000), as well as the desired FDR (default: 0.05; *see* **Note 6**), and click "OK" to run.

A results worksheet will be generated reporting all genes with significantly correlated DNA copy number and gene-expression (DR-Correlate), or all genes with statistically significant copy number and gene-expression differences between group-1 and group-2 (DR-SAM). Each gene's information, score, and associated FDR will be listed, ordered by significance level. Additionally, a dialog box will appear with options for generating plots and heatmaps of the significant genes.

4. Notes

1. The software tool Cluster, freely available from the Eisen Lab (http://rana.lbl.gov/), is a useful tool for normalizing and centering microarray data. For data derived from Affymetrix GeneChip or gene-expression arrays, the third-party software tool dChip is an alternative for normalization and probe set filtering, and is freely available at http://www.dchip.org/

2. Several different algorithms have been developed to call gains and losses from DNA copy number data, and the performance of each may depend on the dataset analyzed. Lai et al. and Willenbrock et al. provide reviews of the most widely used algorithms to date and report the results of a comparative analysis *(28, 29)* (*also see* **Chapter 8** in this textbook). Finally, the web-based tool CGHweb is an excellent resource to compare new algorithms with those already developed *(30)*.

3. DR-Integrator runs as a plug-in for Microsoft Excel on Windows XP/2000/Vista. To install DR-Integrator, double-click the installation file after downloading and note the directory in which the files are installed. Start MS Excel and select Add-Ins from the Tools menu. Click "Browse" and navigate to the installation folder. Select the DR-Integrator.xla file located in the Addin folder, and click "OK". Make sure the DR-Integrator package is checked in the Add-Ins list and click "OK". The DR-Integrator controller buttons should appear in the toolbar. To run an analysis, select all the data in an Excel spreadsheet (including gene and sample labels), and click the "DR-Integrator" button.

4. Selecting the Pearson's correlation option will compute a parametric test of correlation (assumes data is normally distributed) between the DNA copy number and gene-expression values. The Spearman's rank correlation computes a non-parametric test of correlation between the ranks of the DNA copy number and gene-expression values, thereby assessing the direction of correlation, disregarding the magnitudes of the data values in the calculation. Finally, the extreme *t*-test looks for significant expression differences between samples that are copy number outliers (i.e., greatest amplifications vs. greatest deletions). Each statistic of correlation has its merits and could be tried separately.

5. The DR-Integrator analyses require a complete data matrix (i.e., no missing data values), and therefore missing data values are estimated by the k-nearest neighbors imputation method. For each gene i with a missing value, the ten nearest neighbors ($k = 10$ by default) to gene i among samples without a missing value for gene i are identified. The missing data value is imputed using the average of the (non-missing) values for that gene. The checkbox option toggles whether a new worksheet included these imputed values is generated.

6. A FDR of 5% is typically used as the threshold for significance when a single analysis such as DR-Correlate or DR-SAM is performed. However, intersecting multiple gene lists, where each is derived from an analysis that applied a FDR of 5%, often leads to overly conservative results and an elevated false negative rate. Therefore, if the gene lists resulting from DR-Integrator are to be integrated with results from another analysis, a higher FDR would be acceptable (10–20%).

Acknowledgments

The authors would like to thank Robert Tibshirani for helpful discussions in the development of DR-Integrator. K.S. is a Paul & Daisy Soros Fellow and a fellow of the Medical Scientist Training Program.

References

1. Schena, M., Shalon, D., Davis, R. W., and Brown, P. O. (1995) Quantitative monitoring of gene expression patterns with a complementary DNA microarray. Science **270**, 467–470.

2. Calin, G. A., Liu, C. G., Sevignani, C., Ferracin, M., Felli, N., Dumitru, C. D., et al. (2004) MicroRNA profiling reveals distinct signatures in B cell chronic lymphocytic leukemias. Proc Natl Acad Sci USA **101**, 11755–11760.

3. Haab, B. B., Dunham, M. J., and Brown, P. O. (2001) Protein microarrays for highly parallel detection and quantitation of specific proteins and antibodies in complex solutions. Genome Biol **2**, RESEARCH0004.

4. Pinkel, D., Segraves, R., Sudar, D., Clark, S., Poole, I., Kowbel, D., et al. (1998) High resolution analysis of DNA copy number variation using comparative genomic hybridization to microarrays. Nat Genet **20**, 207–211.

5. Pollack, J. R., Perou, C. M., Alizadeh, A. A., Eisen, M. B., Pergamenschikov, A., Williams, C. F., et al. (1999) Genome-wide analysis of DNA copy-number changes using cDNA microarrays. Nat Genet **23**, 41–46.

6. Kennedy, G. C., Matsuzaki, H., Dong, S., Liu, W. M., Huang, J., Liu, G., et al. (2003) Large-scale genotyping of complex DNA. Nat Biotechnol **21**, 1233–1237.

7. Yan, P. S., Chen, C. M., Shi, H., Rahmatpanah, F., Wei, S. H., Caldwell, C. W., et al. (2001) Dissecting complex epigenetic alterations in breast cancer using CpG island microarrays. Cancer Res **61**, 8375–8380.

8. Weinmann, A. S., Yan, P. S., Oberley, M. J., Huang, T. H., and Farnham, P. J. (2002) Isolating human transcription factor targets by coupling chromatin immunoprecipitation and CpG island microarray analysis. Genes Dev **16**, 235–244.

9. Silva, J. M., Mizuno, H., Brady, A., Lucito, R., and Hannon, G. J. (2004) RNA interference microarrays: high-throughput loss-of-function genetics in mammalian cells. Proc Natl Acad Sci U S A **101**, 6548–6552.

10. Segal, E., Friedman, N., Koller, D., and Regev, A. (2004) A module map showing conditional activity of expression modules in cancer. Nat Genet **36**, 1090–1098.

11. Tan, K., Tegner, J., and Ravasi, T. (2008) Integrated approaches to uncovering transcription regulatory networks in mammalian cells. Genomics **91**, 219–231.

12. Lee, S. I., Pe'er, D., Dudley, A. M., Church, G. M., and Koller, D. (2006) Identifying regulatory mechanisms using individual variation reveals key role for chromatin modification. Proc Natl Acad Sci USA **103**, 14062–14067.

13. Carroll, J. S., Meyer, C. A., Song, J., Li, W., Geistlinger, T. R., Eeckhoute, J., et al. (2006) Genome-wide analysis of estrogen receptor binding sites. Nat Genet **38**, 1289–1297.

14. Yu, J., Cao, Q., Mehra, R., Laxman, B., Tomlins, S. A., Creighton, C. J., et al. (2007) Integrative genomics analysis reveals silencing of beta-adrenergic signaling by polycomb in prostate cancer. Cancer Cell **12**, 419–431.

15. Pollack, J. R., Sorlie, T., Perou, C. M., Rees, C. A., Jeffrey, S. S., Lonning, P. E., et al. (2002) Microarray analysis reveals a major direct role of DNA copy number alteration in the transcriptional program of human breast tumors. Proc Natl Acad Sci USA **99**, 12963–12968.

16. Garraway, L. A., Widlund, H. R., Rubin, M. A., Getz, G., Berger, A. J., Ramaswamy, S., et al. (2005) Integrative genomic analyses identify MITF as a lineage survival oncogene amplified in malignant melanoma. Nature **436**, 117–122.

17. Kwei, K. A., Kim, Y. H., Girard, L., Kao, J., Pacyna-Gengelbach, M., Salari, K., et al. (2008) Genomic profiling identifies TITF1 as a lineage-specific oncogene amplified in lung cancer. Oncogene **27**, 3635–3640.

18. Kwei, K. A., Bashyam, M. D., Kao, J., Ratheesh, R., Reddy, E. C., Kim, Y. H., et al. (2008) Genomic profiling identifies GATA6 as a candidate oncogene amplified in pancreatobiliary cancer. PLoS Genet **4**, e1000081.

19. Zardo, G., Tiirikainen, M. I., Hong, C., Misra, A., Feuerstein, B. G., Volik, S., et al. (2002) Integrated genomic and epigenomic analyses pinpoint biallelic gene inactivation in tumors. Nat Genet **32**, 453–458.

20. Zender, L., Spector, M. S., Xue, W., Flemming, P., Cordon-Cardo, C., Silke, J., et al. (2006) Identification and validation of oncogenes in liver cancer using an integrative oncogenomic approach. Cell **125**, 1253–1267.

21. Kim, M., Gans, J. D., Nogueira, C., Wang, A., Paik, J. H., Feng, B., et al. (2006) Comparative oncogenomics identifies NEDD9 as a melanoma metastasis gene. Cell **125**, 1269–1281.

22. Maser, R. S., Choudhury, B., Campbell, P. J., Feng, B., Wong, K. K., Protopopov, A., et al. (2007) Chromosomally unstable mouse tumours have genomic alterations similar to diverse human cancers. Nature **447**, 966–971.

23. Butte, A. J. (2004) *Exploring Genomic Medicine Using Integrative Biology.* Massachusetts Institute of Technology: Cambridge, MA.

24. Liu, F., Park, P. J., Lai, W., Maher, E., Chakravarti, A., Durso, L., et al. (2006) A genome-wide screen reveals functional gene clusters in the cancer genome and identifies EphA2 as a mitogen in glioblastoma. Cancer Res **66**, 10815–10823.

25. Weir, B. A., Woo, M. S., Getz, G., Perner, S., Ding, L., Beroukhim, R., et al. (2007) Characterizing the cancer genome in lung adenocarcinoma. Nature **450**, 893–898.

26. Salari, K., Tibshirani, R., and Pollack, J. R. DR-Integrator: an integrative analysis tool for DNA copy number and RNA expression microarray data. Manuscript in preparation.

27. Tibshirani, R., and Wang, P. (2008) Spatial smoothing and hot spot detection for CGH data using the fused lasso. Biostatistics **9**, 18–29.

28. Lai, W. R., Johnson, M. D., Kucherlapati, R., and Park, P. J. (2005) Comparative analysis of algorithms for identifying amplifications and deletions in array CGH data. Bioinformatics **21**, 3763–3770.

29. Willenbrock, H., and Fridlyand, J. (2005) A comparison study: applying segmentation to array CGH data for downstream analyses. Bioinformatics **21**, 4084–4091.

30. Lai, W., Choudhary, V., and Park, P. J. (2008) CGHweb: a tool for comparing DNA copy number segmentations from multiple algorithms. Bioinformatics **24**, 1014–1015.

INDEX

Note: The letters 'f', 't' and 'n' following the locators refer to figure, table and note number respectively